UNLOCKING
The MYSTERIES of CREATION

by

Dennis R. Petersen, B.S., M.A.

This textbook belongs to:

Name

Address

City

CREATION RESOURCE FOUNDATION

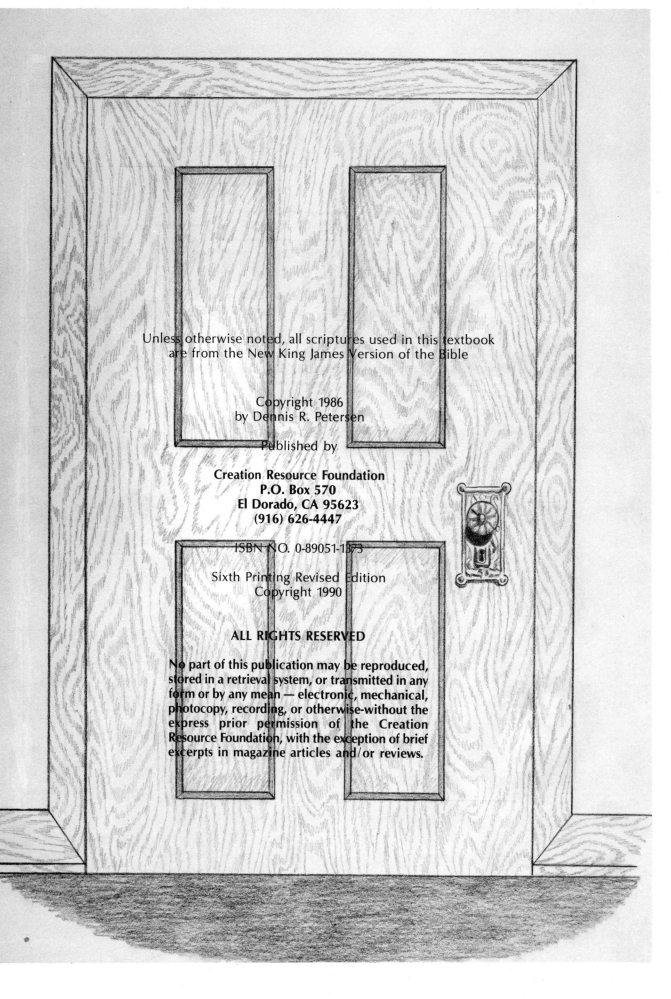

Dedication and Acknowledgments

How can one properly dedicate a new book? In Colossians 3:23 we are told: "Whatever you do, do your work heartily (literally, from the soul), as for the Lord and not for men." When you've put your very heart into a project like this, there is surely no more worthy way to dedicate it than to submit it to the King of all kings, Jesus Christ of Nazareth, in whom are all treasures of wisdom and knowledge. To the end that this book will exalt Him and help human souls everywhere grow in the knowledge of Him, I humbly dedicate this work.

There are many wonderful people whom my Lord has graciously allowed to bless my life in many ways. This book, in fact my entire life's work now, could never have been without the encouragement, confidence, and insight of one of God's choice servants, Reverend Glen S. McLean. I thank God for allowing me to be so deeply affected by his ministry, and I pray that God's investment in my life through him will produce a bountiful effect through this book.

Our personal Lord and Creator uses many ingredients to produce His desired fruit in each of us. Recently, He has prompted many dear men and women to encourage me to write this book. As I have shared these insights through seminars in various communities, pastors and others have prayed with me for this book you are now holding in your hands to become a reality. Since it takes people to accomplish a vision, God has miraculously brought His obedient servants together to produce this work. The involvement of each person is a small miracle story in itself. For Pastor Terry Edwards and Evangelist Francis Anfuso of Christian Equippers International, I am deeply grateful. Through their caring fellowship the Lord has enriched my limited efforts. In the person of art co-ordinator, Ted Evans, the Lord has blessed this presentation with energetic artistic appeal. To artists Don Harris, Steve and Ariane Leonhard, I am also indebted for their fine work. Thanks to the tireless efforts of Warren and Martha Dayton, we were able to complete the major artistic part of the book. Their special talent and gifted ministry is deeply appreciated. To Brian Jones goes my sincere gratitude for his work of computerizing the text. With the sharp eyes of Justine Wolters and Gary Beasley, the manuscript was proofed carefully; how I thank God for their patience and time. Others have volunteered their precious time too. I thank them all, and most of all, I want to thank my wife, Vi, for patiently enduring the hardships of mothering our four precious children while Daddy is in his office most nights, "cranking out the book."

Dennis Petersen
El Dorado, California
Spring 1988

"Do you not Know? Have you not heard? The Everlasting God, The Lord, the Creator of the ends of the Earth does not become weary or tired. His understanding is inscrutable. He gives strength to the weary, and to him who lacks might He increases power."

Much appreciation also goes to artist Kris Westbeld, for the many new pencil sketches she prepared for this edition.

FOREWORD

On many occasions I have had discussions with people who presented arguments supporting "Evolution" that appeared to be scientifically and intellectually sound. Though I firmly believed in the Bible's account of the creation of the universe, I felt unable to adequately represent its teachings from a purely rational perspective. Consequently, my response would be, "I don't know exactly how everything came together, but I know God is the Creator of all things."

My faith in Jesus Christ and the Word of God protected me from being deceived by "the doctrines of men", but this was at best a defensive position. I longed for the needed wisdom to uproot the deceptive lies of the enemy.

Since my primary responsibilities as an "Equipping Evangelist" have not involved a Biblical and Scientific study of the Creation vs. Evolution controversy, I have hoped others would champion this cause. How elated I was to meet Dennis Petersen and attend his "Unlocking the Mysteries of Creation" Seminar. At last someone had spent the necessary time simplifying the wealth of factual data needed to substantiate the Biblical account of the creation.

You will find the information in this textbook to be Bible-based, thoroughly researched, and yet presented in a concise, easily understood format. Truly this is an equipping textbook you will use for many years to come.

The staff at Christian Equippers International is honored to endorse a man of God like Dennis Petersen, who has invested over 12 years of his life preparing this invaluable tool for the Body of Christ.

As I have considered its contents, I believe I have finally been given the ammunition I so desperately needed. I am convinced that because of the truth and wisdom contained in this manual, each of us will be better able "to contend earnestly for the faith which was once for all delivered to the saints." (Jude 3)

Equipping for Jesus,
Francis Anfuso
Vice President,
Christian Equippers International

Preface

The book you are about to experience is likely to be much different from others you have read. This is intentional. Just leafing through this volume will activate your natural inclination to explore and discover the world around you.

As a fellow explorer I find that most people have a God-given curiosity to investigate their world. But how we view or interpret our world is a phenomenon that develops throughout our lifetime and is based on the ideas and concepts that are repeatedly reinforced by what we focus our attention on.

The popular modern concept of the world around us frequently ignores the power and concern of the Living Creator. The Bible is often treated as a concoction of religion and mythology. Miraculous events are regarded as allegorical and impossible. At the same time there is now an epidemic fascination with perverted supernatural things.

Solomon wisely wrote that the true origin of knowledge is found in our personal listening to God. The Apostle Paul asserted that all the treasures of wisdom and knowledge are found in the Savior, Jesus Christ. The reason there are so many "mysteries" about our world is because of our lack of perfect knowledge. Only The Creator has that. Science or philosophy alone cannot possibly have all the answers. In fact, if God is left out of one's "world-view" he will have an inferior grasp of the real world.

The Creator is the original scientist. In fact, He implores us to "prove all things," and to "speak to the Earth." If our mind is truly willing to discover truth, we will find that God's Word, the Bible, will guide us to unlock the "mysteries" of creation.

This book, as with the seminar from which it comes, is not intended to get deep into technical detail. It is for the non-professional "layman." Before God we are all "laymen" and none of us know very much. My desire is to help re-orient people's thinking toward God and His Word. We do not need to be dogmatic about "interpretations": they should always be open to review anyway. What is important is that our inner man is open to hearing from God and responding to His Word. The accuracy of the Bible is an amazing reality. It unveils the deceptions that can enslave us. And we all need that freedom, don't we?

I am indebted to many brilliant "explorers" of the real world. I encourage you to make use of the references listed in each chapter of this book. If we are diligent to study and dig out the truth we will be acceptable workmen to God who are unashamed of our cherished relationship with Him. May this be a stepping stone to your further discovery of the wonderful Lord of Life, who Himself is the One with the key to unlock the mysteries of creation.

Dennis R. Petersen
Spring 1986
El Dorado, California

Table Of Contents

SESSION 1

UNLOCKING THE MYSTERIES OF THE EARLY EARTH

"YOU ALONE ARE THE LORD;

YOU HAVE MADE HEAVEN, THE HEAVEN OF HEAVENS,

WITH ALL THEIR HOST,

THE EARTH AND ALL THINGS ON IT,

THE SEAS AND ALL THAT IS IN THEM,

AND YOU PRESERVE THEM ALL.

THE HOST OF HEAVEN WORSHIPS YOU."

Nehemiah 9:6

Contents Of Section One

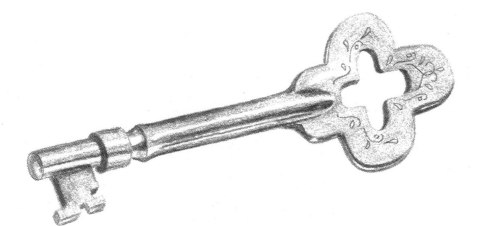

"For in six days the LORD made heaven and earth,
the sea, and all that is in them,
and rested the seventh day..."
Exodus 20:11

Day 1	Time begins	verse 1
	Heaven and Earth created	verse 1
	God's Spirit hovers over waters	verse 2
	Light called to "be"	verse 3
Day 2	Expanse (firmament) made	verse 7
	Waters separated below and above the expanse	verse 7
Day 3	Waters below the expanse gathered to become seas	verse 9, 10
	Dry land appears and is called Earth	verse 9, 10
	Earth brings forth vegetation	verse 12
Day 4	Lights made in the expanse	verse 14, 15
	Sun and Moon made	verse 16
	Stars made	verse 16
	Sun, Moon and stars are placed in the expanse	verse 17
Day 5	Great sea monsters created	verse 21
	All aquatic life created	verse 21
	Every winged bird created	verse 21
Day 6	Beasts of the Earth made	verse 25
	Cattle made	verse 25
	Everything that creeps on the dry ground made	verse 25
	Man created	verse 27
	The rule over every living thing is given to man	verse 28
	Every plant on Earth's surface is given for food	verse 29

SOURCE: Genesis, Chapter 1

In the beginning God created the heavens and the Earth . . .

Then God saw everything that He had made and indeed it was very good. So the evening and the morning were the sixth day.

Genesis 1:31

THE APOSTLE PAUL-Romans 1:18-21

There are *"men who hold the truth"* (KJV) (NAS - *"suppress the truth"*).

God's wrath is upon them.

Why?

Because there are facts which can be *"known about God."* (v. 19)

Paul says God has shown it to them. What has He shown to all men?

"For since the creation of the world His invisible attributes, His eternal power and divine nature, have been clearly seen, being understood through what has been made, so that they are without excuse." (v. 20 NAS)

What has been clearly seen?

GOD'S VERY NATURE!

And how is THAT revealed?

THROUGH THE THINGS THAT ARE MADE!

The created physical world around us points us to God! That's why all men are without excuse for ignoring God.

But what happens when men choose to ignore God? They still have that probing question:

WHERE DID I COME FROM?

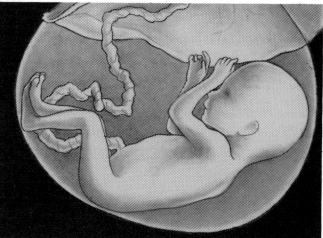

So Paul tells us what we should expect to find.

"But they became vain in their imaginations." (v.21 KJV) (the NAS says: *"futile in their speculations"*)

Note: Have you ever noticed the utter futility of the reasoning of men who ignore God?

Now let's consider another idea here. When you know your God IS big enough to CREATE, do you think that your FAITH in Him is enlarged?

"Ah Lord God, Behold, You have made the heavens and the Earth by Your great power and outstretched arm! Nothing is too hard for You."

Jeremiah 32:17

How did the infant church pray in confidence to God?

"Lord, You are God who made heaven and Earth, and the sea, and all that is in them . . . "

Acts 4:24

(Did they leave anything out?) They finished by asking God to give His bondservants boldness and confidence to speak while God did signs and wonders in the name of Jesus.

"O Lord how manifold are Your works! In wisdom You have made them all! The Earth is full of Your possessions."
Psalm 104:24

CREATION AND OUR FAITH

The wonderful world of nature is the creation all around us. It was designed and spoken into existence by God's very Word. He is the author of "science."

*"In **Him** are hidden all the treasures of wisdom and knowledge."* (or philosophy and science)
Colossians 2:3

The creation, so frequently used to describe God's power in both Old and New Testaments, shows us that understanding the principles of God's miraculous creation epoch is essential to our faith.

"Through faith we understand that the worlds were framed by the word of God, so that things which are seen were not made of things which do appear."
Hebrews 11:3

CREATION AND CHRIST

Christ himself repeatedly acknowledged the truth of ALL of the Old Testament.

"If you believe not his (Moses') writings, how shall you believe my words?"
John 5:47

Moses penned the creation account and Jesus makes direct reference to it in Matt. 19:4 when speaking about marriage and the creation by God of male and female at the beginning.

"If they do not hear Moses and the prophets, neither will they be persuaded, though one rise from the dead."
Luke 16:31

Think of the implications in sharing the good news!

"O foolish ones and slow of heart to believe in ALL that the prophets have spoken."
Luke 24:25

(Who was Israel's first prophet?)

THE CREATION FOUNDATION

● Genesis - God's first revelation of the Bible - begins with it.

● The Gospel of John begins with it.

● Romans - the complete gospel - makes the knowledge of it the basis for judgement.

● Paul's Mars Hill discourse begins with Creator God. (Acts 17:22ff)

● Colossians begins with it.

● Hebrews establishes creation as its first basic doctrine.

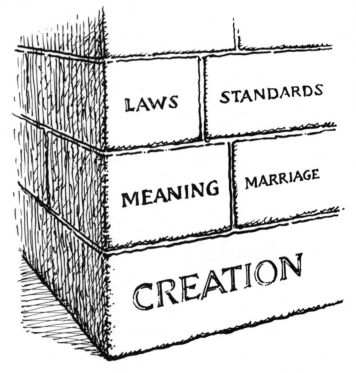

WHY SO IMPORTANT?

Satan desires to undermine God's expressed account of creation. By denying creation it is easier to deny God's plan of salvation and then the truth of all God's word.

If God is subtly removed from people's minds as their Creator, then it is simple for them to disregard Him as their Redeemer.

Our minds are the most intricate computers imaginable. When we look at anything in our world we can only evaluate it on the basis of what we've already learned or heard, and accepted.

Have you ever heard this phrase from computer experts?

"GARBAGE IN - GARBAGE OUT"

Misinformation in our computer sadly affects how we think and believe about everything. Do you think the Devil knows that? Of course he does. Have you noticed how the world of knowledge is such a playground for the Devil?

Someone has observed: "How hard, if not impossible, it is for the heart to accept what the mind rejects!" Does our mental conditioning affect our spiritual receptivity?

IF ANY PART OF GOD'S WORD IS RATIONALIZED AS WRONG, THEN HOW CAN WE BELIEVE THE REST OF IT?

WHAT IS SCIENCE?

THINK!

Isn't science supposed to deal with reality? Isn't it the study of FACTS as they really are? Pure science analyzes THE REAL WORLD.

Wait! How much of what is popularly called science is loaded with speculation and guesswork?

Are we really willing to program OUR COMPUTERS with PURE KNOWLEDGE? Can such knowledge even be found?

WHERE IS THE TRUE KNOWLEDGE TO BE FOUND?

The fear (reverence) of the LORD is the beginning of knowledge.

Proverbs 1:7

The whole structure of the Bible rests on the foundation of Genesis. Why do you think there have been such insidious attacks on things like Biblical creation, the global flood, and other supernatural events of Genesis?

DID YOU KNOW?

- Though some say the Bible is not a science text, the scriptures speak authoritatively in matters of science and history!

- All investigation has shown the Bible to be scientifically ACCURATE!

- Of the two places where the word "science" appears in the Bible note this caution: *"Avoid profane and vain babblings, and oppositions of science, FALSELY so called."*

1 Timothy 6:20

"For as the heavens are higher than the Earth, so are My ways higher than your ways, and My thoughts than your thoughts."

Isaiah 55:9

SCALE OF WISDOM & TRUTH

DOES GOD REALLY EXPECT US TO TAKE HIS WORD AT FACE VALUE?

By the study that God encourages you can demonstrate that the Bible is the Book of Truth. No amount of "Scientific theory" can ever explain away scriptural fact!

The Word of God is:

1. Enduring

"Forever O Lord, Your word is settled in heaven."

Psalm 119:89

"Heaven and Earth shall pass away but My words shall not pass away."

Matthew 24:35 (KJV)

2. Inerrant

"All scripture is God-breathed."

2 Timothy 3:16 (NIV)

"Every word of God is pure . . . Do not add unto His words . . . "

Proverbs 30:5-6

" . . . scripture cannot be broken."

John 10:35

"Let God be true, though every man be found a liar."

Romans 3:4 (NAS)

3. Clear

God's wisdom is . . . "all . . . righteousness; nothing crooked or perverse is in it . . . " All His words are *"plain . . . and right to those who find knowledge."*

Proverbs 8:8-9

WHAT CAUSES OUR LACK OF UNDERSTANDING?

You really only know what you have been taught (by observation or interpretation). If that is built on error and fallacy, then your conclusions will be distorted until some re-evaluation takes place.

1. Ignorance of God's word and His power

Matt. 22:29

2. Natural (unspiritual) thinking.

1 Cor. 2:14

3. Attraction to fables

2 Tim. 3:7ff, 4:3-4

4. Willful ignorance

2 Pet. 3:4-5

5. Suppression of truth

Romans 1:18 (NAS)

Note: What's the outcome of a foolish premise?

Eccl. 10:12-13 (Living Bible)

"DOES NOT THE EAR TEST WORDS?"

Job 12:11

Our failure comes in not examining closely the foolish premises and conclusions of so-called men of science.

Unlocking the Mysteries of **CREATION**

What Should Our Attitude Be As We Study The Creation Around Us?

"SPEAK TO THE EARTH AND IT SHALL TEACH THEE . . . " Job 12:8

At the outset of our study it is important to be especially aware of our attitude. How do we approach the objects of our study?

1. FAITH

"Through faith we understand that the worlds were framed by the word of God . . . "

Hebrews 11:3

Which comes first: the understanding or the faith? Do you remember how obscure the Bible seemed before you believed on Jesus Christ and were born again by faith?

Also see Luke 24:25

2. CONFIDENCE

"Prove all things, hold fast what is good."

1 Thessalonians 5:21

GOD ISN'T AFRAID OF THE FACTS! NEITHER SHOULD YOU BE.

3. INSIGHT

"We wrestle not against flesh and blood, but against . . . spiritual wickedness in high places."

Ephesians 6:12

Every conflict we encounter has an unseen dimension behind it. Winning an argument should never be our objective. We must be spiritually sensitive and prayerful always if our communication is to have any truly meaningful effect.

" . . . the natural man cannot understand the things of the Spirit . . . but the spiritual man judges all things."

1 Corinthians 2:14-15

In the Amplified Bible it says the spiritual man **appraises all things *(examines, investigates, inquires into, questions, and discerns). Nothing escapes his notice.***

The creation by God is a very spiritual matter. It involves *physical* things indeed. But all of it was created by God who is *spirit*. It deserves being investigated.

4. OPENNESS

"The entrance of Thy words giveth light; it giveth understanding unto the simple."

Psalm 119:130

You know that, compared to God, we just don't know too much do we? Yet the wisdom of Almighty God is available to anyone who humbly recognizes how imperfect his human insights are.

You'll like how Proverbs 8:9 reads in the Living Bible:

"My words are plain and clear to anyone with half a mind if it is only open."

"WHO AMONG ALL THESE DOES NOT KNOW THAT THE HAND OF THE LORD HAS DONE THIS? IN HIS HAND IS THE LIFE OF EVERY LIVING THING AND THE BREATH OF ALL MANKIND."

Job 12:9-10

What Really Happened "In The Beginning" According To The Bible?

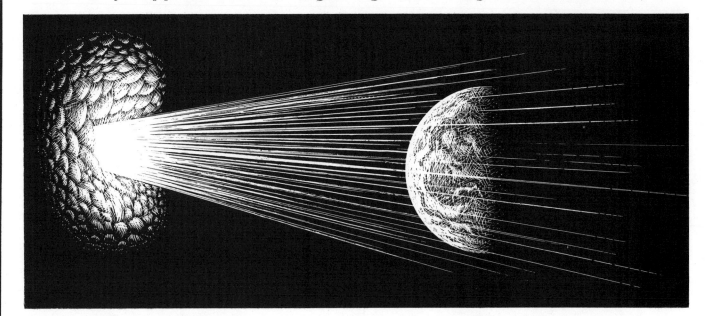

WHEN DID TIME BEGIN?

"In the beginning God . . . "

Only God himself is eternal. What did He do? He "created the heavens and the Earth." What is meant here?

When you begin a construction project, what do you start with?

RAW MATERIALS of course.

ATOMS: What are they?

The building blocks of all physical reality, they are composed essentially of three things:

NOTICE the words of Genesis 1:1

In the beginning (at the outset)

"GOD CREATED THE HEAVENS"

SHAMAYIM (Hebrew for *"Heavens"*) simply means STRETCHED OUT SPACE!

THINK! God even had to make the empty space in which to put everything.

Then what?

"GOD CREATED . . . THE EARTH"

ERETS (Hebrew for *"earth"*) simply means the dirt or MATTER!

What was the condition of these RAW MATERIALS in the beginning?

"WITHOUT FORM AND VOID"

THINK! When you take the blocks and build something, THEN you have "form." Before that the MATTER was without form. The Earth was truly EMPTY (void).

And then, in that same creative instant, God said:

"LET THERE BE LIGHT!"

Light includes the entire electromagnetic spectrum, not just the narrow band of color we "see." From short wave gamma rays, to long radio waves, all energy forms are accounted for here.

So what do we now have? Is it not the third aspect of all physical things?

ENERGY

At the instant electrons began their atomic movement, then processes were occurring in TIME.

THINK! When physical matter ceases to be, time shall be no more.

In a very simple, yet profoundly scientific way, the Genesis account of God's first creative act logically defines the basis of all physical reality:

MAGNETS AND MIRACLES

In the nucleus of all atoms is a scientific mystery! All those positive-charged protons stick together! But you take two bar magnets and try to push the two positive ends so they touch each other. What happens?

They repel each other, don't they?"

Because protons adhere closely in the nucleus of every atom, evolutionists have theorized there must be something they call "gluons" to hold it all together. It seems a miracle of "science" that the structure of every atom in the universe doesn't fly apart. What does the Bible say?"

*"...all things were created by Him (Jesus) and for Him; And He is before all things, **and by Him all things consist** (or hold together)."*

Colossians 1:16,17

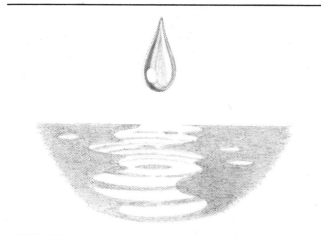

THE CREATION OF WATER

First mention:

In Genesis 1:2 we read of God's Spirit moving (or hovering) over the face of the waters. But no mention is made of how the water got there. The separation of waters by a firmament in verse 6 presumes that the waters have already been created.

It's interesting to notice that the Hebrew word for waters is "MAYIM." The first instance of the word "mayim" is in a compounded form with the word "SHAM," meaning "there" or "in it." It is thus found as the word "SHAMAYIM" and is translated as "heavens" in Genesis 1:1. It appears there is something about the expanse of "heavens" which inherently includes water.

Scientists are learning that there is water out in the emptiness of interstellar space. (Nat'l Geo. 5/74 p. 625)

What are "waters above the heavens" in Psalm 148:4?

A Parallel In Uniqueness

Water is one of the most primary of physical phenomena, yet it defies normal chemical theory. The chemical union of two gases, it is not easily produced or broken down. Virtually a liquid mineral, it is a misfit to other compounds, expanding when it freezes, enabling it to float.

Water is absolutely essential in liquid form for all the key systems of life. Circulation, digestion, reproduction, and respiration are all dependent on liquid water.

WITHOUT WATER THERE IS NO LIFE!

Space probes to other planets always search for liquid water to see if life forms could be present. But Earth, the place where God specifically created physical life forms, is the only planet where water is found!

Note: John 6:63 (NAS)

"IT IS THE SPIRIT WHO GIVES LIFE"

Remember the connection of the first Bible reference for water? It's also the first time in the Bible where we find God's Spirit mentioned.

This might not be just a coincidence. Just as water is essential for physical life, so the Spirit is essential for our spiritual life. The fact that God's Spirit interacted with water at the beginning seems to amplify the uniqueness of both in respect to their life-giving qualities.

The Evening And The Morning Were The First Day

WHAT IS LIGHT?

Try as man has, he simply cannot adequately isolate, examine, and define light. Consider:

● God spoke and there was light

verse:

● His word gives light/is light

verse:

Indeed, there is a spiritual complexity here, which even we who walk in the light, will not be able to fully comprehend till that glorious day when we will stand in His very presence, seeing Him just as He is. (1 Jn 3:2)

Besides the known electromagnetic spectrum (of which only a small part is visible) there may very well be realms of light and energy unknown to science.

Until recently it was thought that the speed of light was the ultimate possible speed at 186,282 miles per second. But now incomprehensible things in the universe are being clocked at speeds apparently faster than that.

What happened in that brilliant moment when God said *"Let there be light"*? Being *"very good"* and unaffected by sin's curse, the entire universe might have bathed in the marvelous light of the Creator's presence.

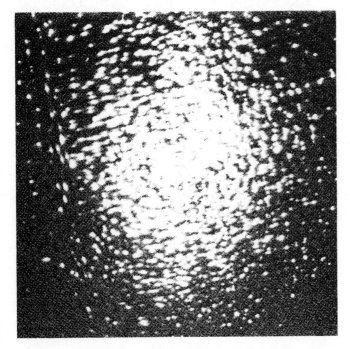

All this so far happened the first day. Naturally the logical question arises:

HOW LONG WERE EACH OF THOSE "DAYS" OF CREATION?

Always in Genesis One the days are set off by the phrase, "the evening and the morning." These are bounds sensible to Moses and us, with nothing more implied than the reality they represent: one revolution of the earth. Could the earth revolve even if the sun is not there yet? Which day did God make the sun? See verses 16 to 19.

In 2 Peter 3:8 God's long-suffering and timelessness is being described in such words. But the context is His judgement, not His creation! It is an expression of attitude and not a declaration of the length of the seven days of the creation week.

The only reason to equate "day" to some long period of time is if we have some outside information. But what does the Bible say about these very days? Look at Exodus 20:8.

"Remember the Sabbath day . . . "

WHY?

See verse 9. *"six days you shall work."* (Hebrew word for day here is "yom")

WHY A SIX DAY WORK WEEK?

See verse 11. *"For in six days (same root word) the Lord made heaven and Earth, the sea, and ALL THAT IS IN THEM."*

DO YOU THINK GOD COULD LITERALLY DO ALL THAT IN SIX DAYS?

THINK! If one of those days God is talking about here really is a thousand years long, have we ever got a long work week before Saturday comes!

Before we try to fit Biblical creation days into any THEORY requiring long ages for Earth's gradual development, look at the problems such a theory causes.

1. Earth existed before all stars!

2. Water covered the Earth before the dry land appeared.

3. Fruit trees existed before fishes.

4. All plants existed before the sun.

5. Birds and whales existed before all land creatures.

6. Man was made before woman!

What about all those stars that are millions of light years away?

WHY WERE THE HEAVENLY BODIES MADE ANYWAY?

God gives six purposes in Genesis One verses 14 and 15. Note the last one: **"TO GIVE LIGHT ON THE EARTH."**

THINK! In God's master plan what is the center of attention?

THE STARS ARE FOR EARTH'S BENEFIT!

DO YOU THINK A CREATOR GOD ACCOMPLISHES HIS PURPOSES WHEN HE SAYS HE DOES?

"By the word of the Lord were the heavens made and all the host of them by the breath of His mouth . . . for He spake, and it was done, He commanded and it stood fast." Psalm 33:6,9

THINK! If your God is powerful enough to create the stars themselves, do you think He just might somehow be able to make their light beams appear instantly on the earth?

With the recent reports of unusual observations in astronomy like bending light in space and the speed of light slowing down, who knows what could have been in the beginning? "HOLD ON," someone may say, "doesn't that create a misleading

APPEARANCE OF AGE?"

THINK! How old was Adam when he rose from the dust? No matter what his stage of physiological development he could not avoid having an appearance of AGE or, perhaps we should say APPEARANCE OF MATURITY!

DID GOD CREATE SEEDS, SEEDLINGS, OR FRUIT TREES WITH MATURE FRUIT ON THEM AND SEED IN THEM? See Genesis 1:12.

"The Lord by wisdom hath founded the Earth; by understanding hath He established the heavens."

Proverbs 3:19

CREATION AND NATURAL LAWS

The acts of creation were governed by God and by the laws He used at creation. Those creative principles are not presently continuing. God finished His creation and deemed it very good before He rested.

Though the laws of creation are not known, we are aware of the laws of "nature" in the created universe. Every natural law operating in the universe is governed by God through principles which He alone established. Gravity, nuclear physics, and the dynamics of light, mass, sound, and energy are all regulated by the NATURAL LAWS OF GOD. Most of them aren't understood; they are only analyzed, and then to just a limited degree.

"He upholds all things by the word of His power."

Hebrews 1:3

Do we tend to overly mystify the things of God? All of God's *"supernatural"* acts are perfectly *"natural"* to HIS nature and power! He is consistent and not whimsical.

THINK! If so-called supernatural events find explanation in the framework of understood natural phenomena, does that lessen the divine miraculous wisdom behind them?

As mankind discovers, analyzes and exploits the creation there is a tendency to make a god of knowledge and science. But for those who recognize the orderliness of God, such discovery is a gateway to the revelation of the only true Master of creation.

"CHRIST HIMSELF, IN WHOM ARE HIDDEN ALL THE TREASURES OF WISDOM AND KNOWLEDGE."

Colossians 2:2-3 (NAS)

Consider:

"The secret things belong to the Lord our God, but the things revealed belong to us and to our sons forever, that we may observe all the words of this law."

Deuteronomy 29:29 (NAS)

CHANGING OUR MIND . . .

It's not easy. We are products of what we are taught, reinforced from childhood. Naturally we get defensive when encountering new concepts. But here is really an opportunity to test our spiritual and scientific maturity.

The idea of "the gradual improvement of all things" has permeated our thinking on just about every subject. Through our study we will discover that this subtle philosophy is based on dogmatic beliefs and not sound scientific integrity at all.

In the public mainstream it seems that those who do not conform to the prevailing dogma are tagged as non-conformist fools. Unfortunately this pressure by what is falsely called science has caused many Christians to reinterpret scriptures in an attempt to harmonize the Bible with evolutionary theory. Thus, explanations are squeezed out of God's word that He never intended. The word "all" is changed to "some," the definitive "did" becomes a progressive "does," and true narrative becomes allegorical poetry.

All this is done to accommodate an idea which does not even have the requirements of a bonafide scientific theory! The whole arrangement is like mythologies of pagan cultures everywhere: you either accept the whole fantasy religiously or you're counted as some kind of kook. But who are the genuine kooks?

We must objectively examine all the evidence. To that must be added all the divine revelation available to us. Then we will have a superior perspective on the real world and the One who designed and energized every delicate detail.

THE ALTERNATIVES AT STAKE

The following is just a beginning of what could be listed of the choices one has to make when he accepts one side or the other. Add others to the list as you think of them.

Divine Creation	Spontaneous Generation
Purposeful Design	Random Accidental Order
Intricate Order	Chaotic mistakes
Infinite God	Infinite Odds
Life from Life	Life from Non-life
God = Creator	Time = Creator
Entropy	Evolution
Catastrophe	Gradualism
True Bible	True Theories
God's Purpose	No Purpose
Absolutes Exist	Everything is Relative
Mutations are Harmful	Mutations are Beneficial
Relatively Young Earth	Extremely Old Earth
Civilization From Start	Slow Development of Civilization
Degenerate Man	Man Getting Better
A Future Hope	Hopelessness

THE REAL CHALLENGE

From the intricate universe of the microscopic cell and "invisible" atom to the unfathomable expanse of space, you look at the creation and you see order, design, and natural law.

The proof is often demanded of a Christian that an infinite God created such precision, such complexity. What's the "intellectual" alternative? Did all creation just happen on its own? Just examine the evidence and see the hallmarks of the Master Designer. Truly, the fool has said there is no God (Psalm 14:1). And only a fool will insist, against all convincing evidence, that the only god is "eternal matter" and the only mechanism of creation is "endless ages of time."

Water . . . Above The Firmament

WAS EARTH'S ENVIRONMENT RADICALLY DIFFERENT AT THE BEGINNING?

"And God made the firmament (or expanse - of the sky) and divided the waters which were under the firmament from the waters which were above the firmament . . . "

Genesis 1:7

Could this "waters above" have been more substantial than the clouds we see now? Perhaps a massive CANOPY of water vapor surrounding the entire globe was located high in the atmosphere.

CAN SUCH A PHENOMENON REALLY EXIST?

FACT: The planet Venus is shielded by a vapor canopy so dense we cannot see through it to the surface of Venus.

FACT: Saturn's moon, Titan, also has vapor canopies completely surrounding it.

From a purely physical viewpoint the idea of a significant water vapor CANOPY is not really far-fetched.

SCRIPTURAL HINTS?

1. Genesis 7:11 At the beginning of the flood, the **"windows of heaven"** were opened. Could this be more than a figurative expression?

2. Genesis 2:5,6 God had **"not sent rain,"** but an ideal watering system on Earth. In fact, no rain is mentioned until the flood, which could be significant.

3. Hebrews 11:7 Noah was warned about **"things not yet seen."** Could the very idea of rain itself have been a new thing to Noah and his contemporaries?

4. Genesis 9:13-14 The rainbow and clouds were especially significant AFTER the flood as a sign of promise. If rain was a new phenomenon then a rainbow also would be an entirely new thing.

5. Genesis 8:22 Could these extremes of seasonal climatic changes have been a totally new experience to Noah after the flood?

GREENHOUSE EFFECT

A protective CANOPY would greatly shelter Earth's environment, resulting in a very productive "greenhouse" effect.

No extremes of hot or cold!

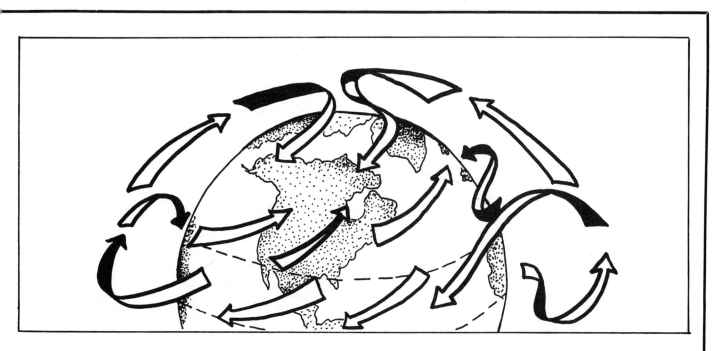

Contrast a CANOPIED environment with present conditions.

Today -

Equatorial zones get the brunt of sun's heat.

Polar regions get little heat. Diverse extremes of temperature prevail.

Today's weather patterns and continual storms are the result of these hot and cold air masses meeting turbulently.

But what would it be like on a sheltered CANOPIED Earth? Diffused sun light and heat would result in:

1. A moderate warm climate prevailing from pole to pole, just like in a greenhouse.

2. No winds and no storms.

3. No encroaching wastelands of desert and ice.

4. A continual growing season.

5. Universal lush vegetation all around the world.

Could a Canopy protect the Earth from harmful radiation?

● The Earth's existing limited "canopies" DO protect to some extent . . .

● With a virtual wall of water to filter or reflect harmful radiation our environment would be extremely well-protected.

WHAT COULD THE RESULTS BE?

● Possible longer life spans

● Possible larger specimens of some plants and animals. [1]

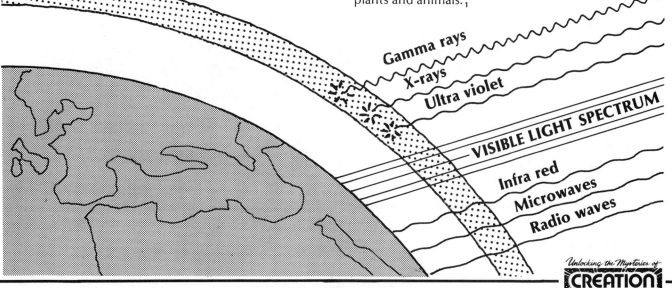

Gamma rays

X-rays

Ultra violet

VISIBLE LIGHT SPECTRUM

Infra red

Microwaves

Radio waves

Unlocking the Mysteries of
CREATION

Does Evidence Tell Of A Very Different Early Earth?

From the fossil remains of plant life found encased in the layers of our planet it is apparent that things grew larger at some time in the past.

Mosses grew **two or three feet** in height instead of just an inch or so as they do today.

Horsetail reeds today ordinarily reach heights of five or six feet. In the past similar plants grew **up to fifty feet tall!**

Dragonflies today have wings that span four inches or so. In the past, however, their wings were up to **three feet across.**

Many other insects thrived in proportionately larger specimens than their modern counterparts. Some cockroaches were as much as a foot long!

Giant animals seem to have been commonplace on Earth in the past. The hornless rhinocerus was about **eighteen feet high and nearly thirty feet long!**

Coiled shellfish today grow up to about eight inches across, but fossilized specimens are displayed in museums that measure over **five feet across!**

Fossil impressions of palm fronds wouldn't be so special if they hadn't been found on **northern Vancouver Island** in Canada. It's obvious that the climate has changed drastically since these tropical plants grew here.

Within the Arctic circle are two very interesting island groups. The New Siberian Islands and the Spitzbergen Islands have both been explored in the last century. Remarkable things have been reported by explorers there. Immense frozen gravel mounds were discovered to have entombed in them *entire fruit trees with the fruit still on them.*[2]

Redwood forests are found buried under massive ice deposits at the **South Polar region.** These towering giants require a very special environment. They are typically found along the northwest coast of the U.S.A. today. But in the past there were giant redwood trees flourishing in many diverse parts of the world as evidenced by many coal and fossil finds.

YES INDEED . . . THE EVIDENCE IS THERE THAT THE ENTIRE EARTH WAS VERY DIFFERENT THAN WHAT WE NOW EXPERIENCE.

Could the environment of the original Earth have been created much more perfectly than what we witness today? Have substantial events changed all that in the history of the Earth since the beginning?

Unlocking the Mysteries of **CREATION**

What Was The Dry Land Like In The Beginning?

WHAT WAS THE CONDITION OF THE WORLD'S LAND MASS?

"the waters were gathered into ONE place."

Genesis 1:9

"the dry LAND He called Earth . . . the waters . . . called seas." (likely including lakes)

Genesis 1:10

THINK! Is it possible God is referring here to only one continental land mass?

We've all noticed on maps how many of the contours of the continental shores seem to fit like so many parts of a big puzzle.

In 1912 a scientist named Alfred Wagner proposed the idea of:

CONTINENTAL DRIFT.

For 50 years his idea was ridiculed by modern scientists. But now it's approved and called the study of the *world's tectonic plates.*

Materialistic scientists of course start the process a supposed 200 million years ago and have the continents moving slowly apart from one another.

But could it just as well have happened suddenly through some massive upheaval of our Earth?

Have you ever noticed the Bible version of this event?

"To Eber were born two sons: the name of one was Peleg; for in his days the Earth was divided . . . "

Genesis 10:25

In this concise statement some awesome events are implied. The word *"Peleg"* literally means **"to divide"** or **"division."** The word "Earth" here speaks plainly of the geophysical Earth rather than any metaphorical expression.

Most interesting is the hidden meaning of the word "divided" in this passage. It literally means to separate or "canal" by water! If this is at all significant it is easy to understand why this father was so motivated to name one of his sons for this spectacular event.

Today the study of the ocean floor confirms that the land masses *HAVE* been ripped apart.

Because this event happened some generations after the flood at the time of the dispersion at Babel, there is reason to believe that the continents did not separate until that time.

WERE VOLCANOES PART OF THE EARTH'S ORIGINAL ENVIRONMENT?

Keep in mind what happens as volcanoes erupt and explode. They exert violent devastation on the surrounding ecological systems. Death is inevitable!

The Bible indicates plainly that the original creation was all **"very good."** (Genesis 1:31)

A number of Bible passages imply strongly that the curse of death did not begin until after Adam's fall in rebellion to God.

Was the original Earth created to be a peaceful, wonderful place for man and animals to thrive in God's blessing?

EARTH'S MAJESTIC MOUNTAINS

The mountains that we know now were non-existent in the landscape of the original Earth. How can we know this?

All our present mountains are made up of massive flood-laid layers. Many are loaded with evidence of volcanism under water.

Though peaceful looking today, all mountains are the result of violent crustal upheavals. They are typical evidences of the Biblical flood.

Note the description of the origin of the present mountains after the flood in Psalm 104:8.

The Creation Of Plants

THE FIRST SPARK OF LIFE

"And God said, Let the Earth bring forth grass, and herb yielding seed, and the fruit tree yielding fruit after his kind, whose seed is in itself, upon the Earth: and it was so."

Genesis 1:11

THE THIRD DAY OF CREATION

Immediately after God made the dry land appear, on the very same day, He commanded the Earth to bring forth billions of plants of every description to fill the land surface of the planet.

This happened even before there was the sun or any other heavenly bodies. The brilliant light of the presence of the Creator Himself flooded the Earth as He delicately fashioned the vast array of growing things.

THE AWESOME LITTLE SEED

To look at most seeds would hardly give one cause to take any special notice. But within that tiny factory lies the miraculous power to perpetuate life itself.

Every growing thing is programmed to produce its own special seed. Scientists can analyze and take apart the many chemical elements of the seed. But in all of its microscopic complexity no one can build a synthetic seed. And there is no such thing as a simple one!

Every seed is only produced by an amazing pattern of sexual union determined by the special construction of the parent plant.

Think how that seed can just set there for years looking perfectly dead. Kernels of wheat are said to have been taken from Egyptian tombs, having been stored there for thousands of years. And when planted and watered life came forth.

NEVER TAKE A PLANT FOR GRANTED AGAIN!

Ponder the specific job and intricate perfection of every part of the plant. Roots, stem, branches, leaves, flowers, and finally the fruit, each have their own special design and function.

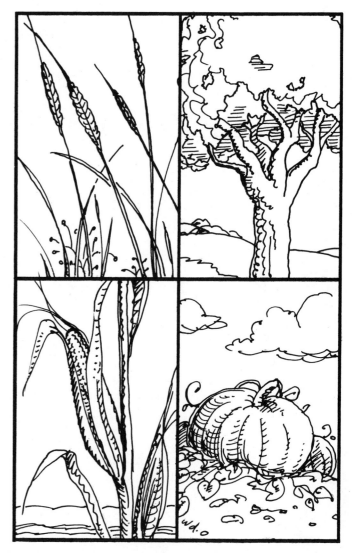

ALL CREATURES AND MAN NEED PLANTS TO SURVIVE

Only plants can take raw elements of the Earth and convert them to food. Animals can't do it. This is the primary reason God made them.

Think how wonderful it is that the Creator did this with such a magnificent display of beauty as the process of growth unfolds.

WHY PLANTS?

"See, I have given you every herb that yields seed which is on the face of all the Earth, and every tree whose fruit yields seed; to you it shall be for food. Also, to every beast of the Earth, to every bird of the air and to everything that creeps on the Earth, in which there

is life, I have given every green herb for food;" and it was so.

Genesis 1:29-30

HOW PRODUCTIVE?

Think how vast a crop can come from a single seed in just a few seasons. A kernel of corn yields hundreds in a few months. A dot-sized poppy seed can produce tens of thousands like itself in one summer.

MIRACLES OF PLANTS

Imagine a factory converting raw dirt plus common water and sunshine into useful and edible products! Besides foods there are textile fibers, lumber, rubber, oils, and innumerable derivatives from plants that are useful to us.

It all happens painlessly in the most efficient manufacturing system imaginable, without even a whisper of noise. There's no complicated machinery to break down. And it's been going on faithfully ever since that third day of creation.

If it weren't for plants we'd all choke to death! They exhale the oxygen we need while we dispose of the carbon dioxide they need. Think of the balance of this system so skillfully conceived by the Creator.

What a desolate place Earth would be without plants. They limit erosion, influence climate, and give shelter for birds and animals. And from the giant sequoia to the colorful fungi they make our world a continual delight of discovery and inspiration.

HOW PRODUCTIVE WAS THE ORIGINAL EARTH?

Consider the real origin of coal. Repeated sedimentary layers extending for miles with no sign of root beneath, coal seams can range from an inch thick to dozens of feet. If a worldwide, year-long, tidal catastrophe really is responsible, then the evidence is there.

It's estimated that the world's coal reserves contain 1.4 times the carbon in the plants that would fill our Earth if it was all as lush as the tropics today![3]

According to the genealogy lists of the Bible in Genesis 5 and 11 the creation week is pinpointed at about 6,000 years ago!

The dominant popular theory today demands that the Earth is at least four and a half or five billion years old.

It's obvious that between these two extremes of thinking there is a very great gulf. In fact, there is so little likelihood of trying to compromise or reconcile these extreme differences that we are forced to the reality that one of these concepts is ridiculously wrong!

WHAT'S THE DIFFERENCE?

One way to visualize the extremes of our choices is to equate one year to the thinness of one page from a typical Bible. If you were to stack up several Bibles to a height about equal with your knee, you'd have about 6,000 pages before you.

Now how many Bibles would you have to stack up to make four and a half billion pages?

The stack would reach at least a hundred and fourteen miles high into the stratosphere.

So you're standing there between your two stacks and you are supposed to chose which one to believe in. Why is it that you are made to feel rather sheepish to admit that you lean toward the Biblical stack of about 6,000 years? Or why is it that you start to arrogantly ridicule anyone who would dare to not agree with your proud billions?

Which conclusion is backed by historical and scientific fact? In all honesty, we need to ask: "What are the origins of each of these ideas?"

GEOCHRONOLOGY

The science of determining the Earth's age is called ''GEO-'' (meaning Earth) ''-CHRONOLOGY'' (i.e. having to do with a time sequence). So how can you clock earth's age?

Scientists are aware of over 70 methods that can give us ideas of Earth's age. We could call these ''GEOLOGIC CLOCKS.'' All of them are based on the obvious reality that natural processes occurring steadily through time produce cumulative and often measurable results. So these studies reveal maximum upper limits for the time these processes have continued.

Only a few of them are portrayed to support billions of years. Those few are loudly publicized to make evolution digestable to the uninformed public, as if to support the long held myth of gradualism.

GRADUALISM

Gradualism is the evolutionary concept that present slow processes made the mountains and land forms. Dynamic large scale catastrophe is ruled out. Stretching present processes over millions of years supposedly accounts for it all. But if present processes can be seen to verify a relatively young earth, how will that affect our understanding and theories about the origin of our Earth and, for that matter, even the universe?

All of the systems we will explore are scientifically known but generally unpublicized and unknown among even many teachers. We have good reason to ask: *"why is this information suppressed?"*

INTERPLANETARY DUST

Did you realize our Earth is regularly gathering dust from the cosmos? So is the moon! Here on the Earth the dust is hardly detectable. Even so, we should expect millions of tons of it has washed into the sea over the last few billion years. But we don't find it.

Since the moon has no erosion but is also accumulating cosmic dust at a regular rate we should discover something.

At present rates NASA experts were expecting a tremendous layer of dust on the moon due to its 4.5 to 5 billion year supposed age. The most conservative estimates were expecting **54 feet of dust on the moon.** Can you imagine landing in a flour sack that deep?

What a surprise when men did finally land on the moon. They found only an eighth *of an inch to three inches of dust!* That much would have taken *fewer than 8,000 years to stack up.*

Is it possible that maybe the moon really hasn't been there any longer than that?[4]

JUVENILE WATER

When volcanoes erupt on the Earth today as much as 20% of the erupted material is water! This water has come from deep beneath the crust of the Earth, where, being under very high pressure, its temperature was extremely hot. This water soars into the atmosphere as steam and soon condenses down as rain.

This water has never been on the surface of the Earth before so it is called *juvenile water.* Each time another volcano erupts there is more water being added to the oceans; water that was never there before.

> **THINK!** What information can be gained from a process like this to tell us something about the beginning of things?

HOW LONG WOULD IT TAKE FOR ALL THE OCEAN WATER OF THE WORLD TO ACCUMULATE FROM VOLCANIC EMISSIONS ALONE?

Scientists have observed volcanoes erupting at the rate of about a dozen each year. Altogether it has been estimated that their total output of juvenile water amounts to roughly one cubic mile.

By simple mathematics we can now easily calculate backwards to find how long it would take to produce all the present water on the Earth.

How much water fills Earth's oceans, lakes, and streams today?

340,000,000 cubic miles!

Figure it out now. At the rate of one new cubic mile of water being added each year it would take 340 million years to completely account for the origin of all Earth's surface water.[5]

What's the implication?

Based on just this one method alone, the logical conclusion is that **there were no oceans at all on the Earth 340 million years ago!**

But wait!

According to the traditional evolutionary chart of Earth's history 340 million years ago was smack in the middle of the **"age of fishes!"**

Do you see the problem?

Keep in mind that the popular idea of the origin of life assumes that the oceans were essentially full of water at least **two thousand million years ago!**

COMETS

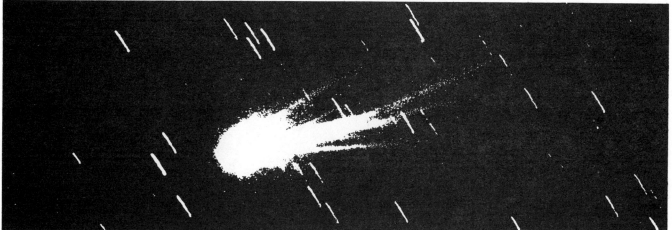

As comets make their circuits around the sun they suffer a fate quite different from the planets in our solar system. Each time they make their swing around the sun they are literally blown apart by the powerful solar wind! On every pass by the sun more of the comet's matter is blasted from its surface to become part of its tail.

Knowing this then it is quite apparent that the comets will eventually disintegrate completely.

How long would that take?

Measuring the observable rate of comet disintegration, scientists realize that **all the short-period comets would be gone in as little as 10,000 years!** Yet there are up to five million comets still orbiting in our solar system.[6]

Because many astronomers are not willing to admit the possibility that this process might indicate that the comets, as well as the solar system itself, were created about 6,000 years ago they are forced to dodge the issue by devising another theory. So they suppose there must be a huge nest of comets far in the outer reaches of the solar system. Every once in awhile some cosmic disturbance supposedly kicks some of these out of the nest.

Has anyone ever seen this nest?

Is there any evidence that such a birth of comets happens periodically?

The answers are no of course. But such explanations must be invoked if one wishes to avoid the glaringly obvious conclusion that the planets and the comets just haven't been there very long.

OIL DEPOSIT PRESSURE

What happens when oil well drillers hit a pocket of oil deep in the Earth? Frequently a gusher goes spouting into the air because of the tremendous pressure trapped below in those sedimentary rocks.

> *THINK!* Even the most dense sedimentary rocks have some degree of porosity. With time, what would happen to all the oil pressure?

Naturally, it would dissipate. And the time it would take is measured in thousands of years, not millions! Findings have revealed tremendous pressures in very deep wells. If those oil deposits had been there for **more than 5,000 years** in some cases there would be **no pressure left!**

The only objective explanation is that these oil deposits were suddenly and catastrophically encased in these flood-produced layers just a few thousand years ago.[7]

What Does The Land Show Us About The Beginning?

EROSION

Consider the present rate of erosion tearing down the world's land mass and filling into the oceans.

The indications of continental geography show that past rates of erosion were much greater than today's rate. But even so, at the present rate of erosion **there should have accumulated at least 30 times more sediment in the ocean than there actually is.** Of course this is based on the assumption that the ocean has been there for at least a billion years.

Even more surprising is the discovery that **all the continents on Earth would be worn down to sea level in just 14 million years.** But there's no evidence for such drastic erosion. The mountains and valleys on Earth appear to have been very recently made. Their sharp angular appearance testifies to their youthfulness.[8]

TOP SOIL

One writer observed that "The soil which sustains life lies in a thin layer of an average depth of seven or eight inches over the face of the land; the Earth beneath it is as dead and sterile as the moon. That thin film is all that stands between man and extinction."

How long does it take for top soil to accumulate?

Scientists estimate that the combination of plant growth, bacterial decay, and erosion produces six inches of top soil in 5,000 to 20,000 years.[9]

If the Earth has been going on about the same as it is today for millions of years one wonders why there isn't a whole lot more top soil than there really is.

Maybe this is just another sign the Earth hasn't been here long.

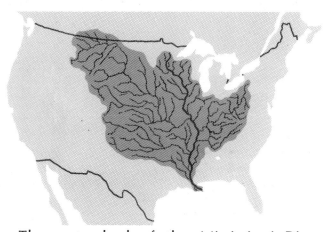

The watershed of the Mississippi River transports a tremendous amount of silt down-

stream to the Gulf of Mexico. Every year the delta deposit is enlarging by a known amount. There is no indication that any of this silt has ever been carried far out to sea.

At the present rate **the entire Mississippi River delta would have accumulated in only 5,000 years.**[10] But science acknowledges that the river has been even bigger in the past.

How could this be? Unless of course the North American continent, and all the other continents for that matter, just haven't been in their present positions any longer than that.

NIAGARA FALLS

This famous waterfall is a manificent example of a GEO-clock that reveals a very young Earth.

Because the rim of the falls is wearing back at a known rate every year, geologists recognize that it has only taken **about 5,000 years** to erode from its original precipice.[11]

CORAL REEFS

The buildup of the calcium carbonate remains of marine creatures in the warm oceans of our world could be accounted for entirely in the few thousand years since the world-wide flood.[12]

STALACTITE GROWTH IN CAVES

If you've toured in a limestone cave you were likely told that the formations of dripstone developed very slowly over a period of more than 100 thousand years. What is the evidence?

Under the Lincoln Memorial in Washington D.C., stalactites had grown to five feet in less than 50 years. Other evidence shows that cave

formations could be easily accounted for in Tens of thousands of years at the most.[14]

IGNEOUS CRUSTAL BUILD-UP

With a dozen volcanoes a year on Earth now there is a steady addition of new igneous rock. There is more volcanic activity on the ocean floor that can not be measured. The observation of our geography shows there have been times in the past of much more intense volcanism.[13]

Conservatively looking at the statistics, the **entire crust** of Earth could have developed without any other process besides volcanism **in only 500 million years.** Could the land have been missing in the Cambrian period? Or is it possible that the dry land was created a relatively short time ago?

WHAT IS THE OLDEST LIVING THING ON EARTH?

Most of us have heard about the antiquity of the giant redwood trees in California. These massive living towers have grown up to more than 300 feet tall. Some of them were already 2,000 years old when Jesus Christ ministered in ancient Galilee.

But there are other plants alive today which date back farther yet.

The twisted and weather-beaten bristlecone pine trees cling stubbornly to life in one of the most hostile environments on Earth where life can exist. In the White Mountains bordering California and Nevada, high in the arid desert, these rare and rugged trees have been growing for about 5,000 years! Their annual growth rings have been studied to give a reasonably accurate idea of their beginnings.[15]

Because of the hardiness of the Bristlecone trees it is fair to say that they will likely go on living for additional thousands of years, barring any catastrophe that would remove them.

The question arises logically: Why don't we find a grove of trees somewhere in the world dating back to 8,000 years, or ten or fifteen thousand years? If trees like this have lived 5,000 years they could have certainly lived longer.

It's almost as though all these trees were planted on a virgin Earth just 5,000 years ago!

The Bible gives a clear historical record pointing to the global flood about 5,000 years ago. With the entire Earth desolated, it stands to reason why the trees of greatest longevity would date back only to that time and no farther.

WHAT DO POPULATION STUDIES SHOW?

One of the most revealing "clocks" deals with the growth of Earth's population down through time. Dr. Henry Morris brilliantly addresses the subject in his book, *Scientific Creationism,* which is highly recommended. Consider the following facts Dr. Morris has gathered.

THINK! If man has been on Earth for a million years why is population explosion only recently becoming a problem?

TODAY - WORLDWIDE

Families average 3.6 children

Annual population grows 2%

FACT: The present population would have been developed from a single family in just 4,000 years if the growth rate were reduced to only 1/2% per year or about an average of only 2 1/2 children per family.

That is a fourth the present rate of growth. It would easily allow for long periods of no growth due to famines and wars.

What does the evolutionary framework have to offer?

With the supposed million-year history of man there would have been an incredible 25,000 generations (at 40 years each). Even more incredible, the final total of people amounts to only the present population of under five billion.

Does this fit the statistical facts?

How big would the population be now if it increased only 1/2% per year for a million years? In other words we would be insisting there be only 2.5 children per family for 25,000 consecutive generations.

The resultant present population would be represented by the number 10 with 2,100 zeros after it!

Obviously, that is impossible since tiny electrons numbering 10 with 130 zeros following, would fill the entire universe!

If a million years of Man's history produced only the present population, how many people would have lived and died in all that time?

It would have been **at least 3,000 billion!** That's at least a couple of dozen graves for every acre on Earth! But ancient bones are extremely rare.[16]

It seems that the facts line up directly with the Biblical model and the acknowledged evidence that human culture can be verified back to less than 5,000 years ago. And that is the point of the global flood.

The observable growth rate of human population on the planet Earth fits much more naturally in a Creation/Global Flood framework than an evolutionary concept of mankind.

THE MAGNETIC FIELD

The phenomenon on Earth that makes directional compasses point to the north and south poles can tell us something about Earth's age. Like everything else in the universe the principle of progressive deterioration (the second law of thermodynamics) is in operation here too.

Physicist Thomas Barnes points out that the Earth's magnetic field has been decaying regularly since it was first measured in 1835. If the half-life of the field is truly what Dr. Barnes has shown from careful measurements the conclusion is that **Earth's magnetic field would have been equal to that of a magnetic star as little as 10,000 years ago.** According to Dr. Barnes, to be consistent with the laws of physics, and assuming the magnetic field has continually weakened, we can only conclude that life on Earth would not have been possible more than 10,000 years ago. [17]

DISSOLVED MINERALS IN THE OCEAN

As world-wide erosional processes continue, a wide variety of minerals is dissolved and carried by rivers into the oceans. The densities of these dissolved minerals is slightly increasing every day.

Because the amounts of these materials can be measured in the river waters and in the oceans, an equation can be calculated to give us an idea of how long it would take, at present rates, for these elements to reach their present densities in the oceans.

Of all the minerals and compounds found in sea water, none of their present concentrations require the assumed evolutionary age for the ocean! The evidence does not suggest the elements have precipitated out of solution. The only conclusion is that the oceans are relatively young. [18]

ATMOSPHERIC HELIUM

The light gas, helium, used to fill balloons, is steadily gathering in the outer reaches of our atmosphere. The total amount there can be measured. One of the sources for it is the constant measurable decay of uranium on the Earth.

If Earth is billions of years old the atmosphere would be saturated with helium to such a degree that there would be up to a million times more helium there than we have now!

Some have suggested that the helium must be escaping into outer space. But actually such escape is impossible. And the indication is that even more helium is being steadily added from the sun.

According to some experts, the helium "clock" insists that the Earth can not be more than 10,000 to 15,000 years old. [19]

THE MOON IS RECEDING

Scientists observe that tidal friction and other things are making the Earth's rotation speed slow down a very tiny amount each year. Though it is not significant enough to make an impact on the Earth even over a few billion years, it does result in another interesting effect. The moon's distance from the Earth is constantly increasing! Two inches a year may not sound like much, but working it back would mean the moon and Earth would be touching only two billion years ago. Of course that's ridiculous. Another way to look at it is this: At the present rate and starting from a realistic distance of separation between the two, if the Earth is 5 billion years old the moon should be out of sight by now![20]

THE SUN IS SHRINKING

A news wire report of March 23, 1980 made the surprising statement: **"The sun's diameter appears to have been decreasing by about one tenth percent per century!"** Scientists have been watching for over a hundred years and the evidence is conclusive. **Every hour the sun is shrinking about five feet.**

Of course five feet an hour isn't much when you consider the sun is nearly a million miles in diameter. (840,000 miles) But what are the implications?

If the sun is shrinking one tenth percent per century then it totals one percent per millennium.

Now if you believe the Earth's age is only 6,000 years there's no real problem. In that time the sun will have shrunk only 6%. But what do you have to contend with if you believe the billions of years idea?

If the sun existed only 100,000 years ago it would have been **double its present diameter.** And only twenty million years ago **the surface of the sun would be touching the Earth!**

As far as researchers can tell this rate of shrinkage has been consistent since the origin of the sun. But astronomers also admit that stars much larger than the present size of our sun burn hotter and faster than the sun. From the pure simple evidence it is clear that life would have been totally impossible on Earth even a million years ago.

Or perhaps the sun and Earth just aren't all that old![21]

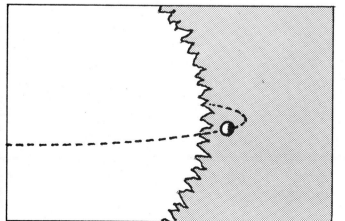

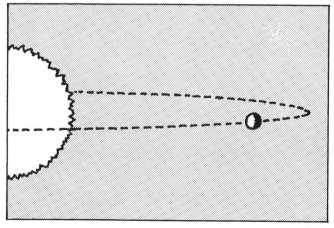

What Can We Learn From The Universe About Its Origin?

Ideas about the origin of the universe are always speculation. Since no man was there to observe and report about it, we must be honest to admit that statements about supposed processes of the past are unconfirmed.

As for present processes, what do they suggest about the dim past of the vastness of space?

NO ASTRONOMER HAS EVER SEEN A STAR BORN, MUCH LESS EVOLVE INTO A MORE COMPLEX STRUCTURE!

All that astronomers have seen is the violent destruction of some stars.

DEGENERATION AND DISINTEGRATION IS CERTAINLY NOT EVOLUTION.

In astronomy, a great deal of importance is placed on "seeing" things. It is extremely faulty "science" to announce a process which cannot be observed. Yet astronomers talk and write often about stars being born as if they were sure of it, and positive that it happened through millions of years of time.

Note: National Geographic, Jan. 1985, page 8 states: " . . . new stars are still being born—as they always will."

THINK! How much integrity is represented in a statement like that?

HOW OLD IS THE UNIVERSE?

Astronomer Harold Slusher reports about a star cluster of four stars in the "trapezium" in the Orion nebula. These four stars are moving away from a common point at a very high speed. Projecting backward in time it appears these stars originated at a common point about 10,000 years ago. The mystery is that there are stars in the cluster which, according to the evolutionary scheme, are millions of years old! No astronomer believes that stars originated before the cluster itself.[22]

HOT STARS

Astronomers have set up a measurement scale by which stars are calibrated according to the amount of energy they emit. Letters are designated for each star type and various colors are detected coming from the various normal star types. Our sun is an average star, labeled as a "G-type." It is yellow with a surface temperature of nearly 10,000 degrees Fahrenheit. Some stars are much, much smaller and considerably cooler. Others are tremendously hotter and larger, emitting energy in the ultraviolet spectrum. The "O-type" stars are over 90,000 degrees with diameters 10,000 TIMES the size of our sun (which has a diameter of 864,000 miles). The "B-type," though not as big, are still monstrous energy burners.

These "O" and "B" stars are radiating more than 100,000 times the energy coming from our sun. Burning down at that rate, and clocking backward, *the entire universe would have been filled with the mass of these stars just a few thousand years ago!* Not millions![23]

MYSTERY?

Some of these "hot stars" are in clusters together with cooler stars, indicating related origins in time. Assuming an advanced age for all of them, Henry Russel wrote in his Astronomy, Volume 1: "Those O and B stars should have been dead for a million years, yet they are present with young ones" (assuming colder stars are younger).

POYNTING ROBERTSON EFFECT

All stars have a gravitational field and pull in particles like gas, dust, and meteors within their range. Stars radiating energy 100,000 times faster than our sun have a spiraling effect, pulling things in all the faster. The unusual thing is that O and B stars are observed to have huge dust clouds surrounding them. If they were very old at all, every particle in close range would have been pulled in by now.[24]

IS THERE EVIDENCE TO DISPROVE THE BIG BANG THEORY?

Did the universe begin with a super-explosion of a small mass 10 - 15 billion years ago? According to demonstrated laws of physics certain things should be expected from a proposed cosmic explosion. Other things would be absolutely impossible in such a framework. The following problems are observed in our solar system and defy the big bang explanation.[25]

1. Suns and planets do not condense from cold clouds of gas and dust.

2. The sun has a very small angular momentum compared to the planets (1/200 th)

3. The system's major angular momentum is in the planets.

4. There are eccentric and even tilted planetary orbits.

5. Uranus and Venus rotate in the opposite direction to the rest.

6. Some of the planets' satellites are also in retrograde motion.

7. There is even distribution of angular momentum among the satellites.

8. Our moon has a lower density than the Earth.

9. The heaviest elements are predominantly in the smaller planets.

"O LORD, Our LORD, how majestic is Thy Name in all the Earth, Who hast displayed (set) Thy splendor above the heavens! When I consider (see) Thy heavens, the work of Thy fingers, the moon and the stars, which Thou hast ordained (appointed, fixed); what is man that Thou dost remember him?"
Psalm 8:1,3,4

"The heavens are the heavens of the LORD, but the Earth He has given to the sons of men."
Psalm 115:16

Are Radiometric Dating Methods Reliable?

During the last few decades the popular press has frequently asserted "millions of years" dates as authoritatively true, based on radiometric tests. The most commonly reported tests are:

1. Potassium Argon (K Ar)
2. Uranium Lead (U Pb)
3. Carbon 14 (C-14)

HOW DO THEY WORK?

All of them are based on several assumptions about the decay process of these unstable elements. For example, radioactive uranium (being unstable) gradually breaks down to form the very stable element lead. Because the decay process is very slow these processes are supposed to be good ways to find ages in the millions and billions of years.

With the exception of carbon 14, these few processes can only be observed in igneous rocks. Thus volcanic rocks (usually) are tested to yield the supposed age of a fossil or artifact found buried beneath them.

The conclusion is that an object buried under a certain layer of rock whose age can be established must be at least as old as the rock, since the rock is assumed to have erupted after the object was imbedded in the underlying stratum.

ARE TEST RESULTS CONSISTENT?

When the Apollo 11 mission brought moon rock and soil samples back the uranium lead tests on them produced 4 different dates:

4.6 billion years

5.4 billion years

4.8 billion years

8.2 billion years

How do we know which figure is correct? Are any of them correct?

According to Science magazine, potassium argon tests on lunar rocks revealed an age of 2.3 billion years. (vol 167, 1/30/70)

WHAT IS THE ACID TEST FOR THESE RADIOMETRIC DATING SYSTEMS?

The best way to test a clock's accuracy is to compare it to a "standard." In other words check with a known reliable source.

CASE HISTORY:

Volcanic lava rocks from Hawaii were subjected to Potassium Argon testing.

Result?

160 million to 3 billion years ago is when these rocks supposedly originated.

Upon further checking it was discovered that the particular lava flow from which these rocks were taken, actually **erupted in the year 1801!**[26]

THINK! Do you suppose there just may be some flaws in the radiometric dating methods?

CASE HISTORY:

Tests were made on volcanic rocks from Russia with results ranging from 50 million to 14.6 billion years. Historical research determined that these very rocks had actually erupted only a few thousand years ago.

CRITICAL QUESTION

When there are discrepancies about a rock's age when the actual age is known, do you suppose there might be similar discrepancies regarding rocks of totally unknown age?

Can we assume these ages are correct when the only tests performed were made on the volcanic ash overlying the artifacts?

IS VOLCANIC ROCK AGE-TESTABLE?

The following familiar discoveries were all "dated" by the radiometric tests on volcanic (ash) material overlying the actual artifacts.

1. Skull "1470" - 2.8 million
   (National Geographic 6/73)

2. "Lucy" - 3 million
   (National Geographic 12/76)

3. Footprints - 3.6 million
   (National Geographic 4/79)

Other famous finds have produced some helpful added data. [Source: Radiocarbon (journal), vol. 11, 1969]

1. Australopithicus - Ethiopia

   ● publicized as 1 to 2 million years old by KAr date for the overlying rocks.

   ● Mammal bones in the same deposit produced a C-14 date of 15,500 years old.

2. Zinjanthropus - Kenya

   ● overlaying volcanic ash gave a KAr age of 2 million years old.

   ● Mammal bones in the same deposit were dated by C-14 at only 10,000 years old.[28]

SURPRISING DISCOVERY

Tree roots were fossilized in moments recently when a high voltage line fell near Grand Prairie, Alberta, Canada (1973).

Scientists at the University of Regina, Saskatchewan, were asked what the results would be if these specimens were dated by KAr. They said the test "would be meaningless; it would indicate an age of millions of years because heat was involved in the petrification process."[27]

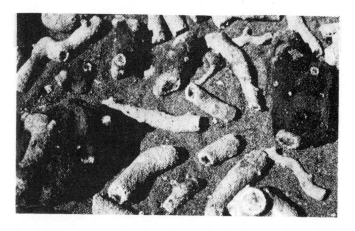

Did you catch that? Heat? What about all the hot volcanic ash we've been finding at other sites?

What is it?

Discovered in 1947 by Willard F. Libby, the Carbon 14 system is used to measure the percentage of unstable carbon 14 isotopes in once living objects. The results of the test are supposed to give the number of years that have passed since the specimen died.

What are its limits?

The "half-life" of C-14 is just 5,730 years. Thus, in five half-lives (29,000 years) very little of the C-14 isotope· would remain in the specimen. An accurate conclusion would be unlikely. ["half-life" is time required for half the unstable element to break down into its "daughter element."] Generally, objects assumed to be older than 50,000 years would not make it into a C-14 test lab.

Is there room for error?

The whole system depends on a steady, unchanging rate of radiation through the atmosphere for at least the last 30,000 years. There are a number of unprovable assumptions in the method.

Let's illustrate in this way. Imagine appearing in a sealed room with nothing but a candle that's been burning there and you are asked to determine how long the candle has been burning.

As you begin to assess the situation you realize the utter impossibility of meeting the challenge. You might think of figuring how long the dripping wax took to accumulate. Or you could get very technical and try to measure the relative amounts of oxygen and carbon dioxide gases in the room compared to outside.

But how can you know if someone had ever opened and re-shut the window? Could any past condition have caused it to burn faster? Was it ever put out and relit?

What things could alter the decay rate of Carbon 14?

1. Atmospheric pollution from
   a. volcanic activity and
   b. industrial burning,
2. Solar activity and changes like
   a. Solar flares and
   b. sunspots,
3. Cosmic radiation from extra-ordinary events in our galaxy like a supernova (explosion of a star), and
4. Meteors or larger cosmic bodies falling to Earth.

An example of how meteors have affected C-14 readings occurred in Siberia on June 30, 1908. Read about the "Riddle of the Great Siberian Explosion" in *Readers Digest*, August 1977. According to reports the carbon 14 measurements of tree rings around the world were greatly altered as a result of the blast, giving inaccurate readings.

How can we measure ANY sample and calculate in the effects of past environmental changes? The C-14 system depends on the idea that there have been NO globally catastrophic events in the past fifty thousand years. *If conditions on Earth were very different in the past, and especially before the flood, then C-14 is nearly worthless, particularly for ages beyond 5,000 years.*[0]

Are there known discrepancies in C-14 dated samples?[29]

1. Mollusks (living) test dated at 2,300 years dead.

2. Mortar from an English castle less than 800 years old- tested at 7,370 years old.

3. Seal skins (fresh) dated at 1,300 years old.

You may wonder how any C-14 dates can be trusted, especially if you do not have another way to test for accuracy.

Do Radiometric Dates Verify The Approved Geologic Charts?

According to the popular traditions of our day, this chart is intended to portray the progressive "Ages" of Earth's history.

SHOWING	DIVISIONS	OF	GEOLOGICAL	TIME	LIFE
	DEVELOPMENT	OF	PLANT	AND ANIMAL	
ERA	PERIOD	ROCKS		DOMINANT LIFE	

DIVISIONS OF GEOLOGICAL TIME — SHOWING DEVELOPMENT OF ROCKS AND DOMINANT PLANT AND ANIMAL LIFE

ERA	Years	PERIOD		
CENOZOIC	2,500,000 Yrs	QUATERNARY	AGE of MAMMALS	PLANTS
	66,000,000 Yrs	TERTIARY		AGE of SEED
MESOZOIC	144,000,000 Yrs	CRETACEOUS	AGE of AMMONITES	MODERN
	208,000,000 Yrs	JURASSIC	AGE of REPTILES	SEED PLANTS
	245,000,000 Yrs	TRIASSIC		AGE of SEED
PALAEOZOIC	286,000,000 Yrs	PERMIAN	AGE of AMPHIBIANS	ANCIENT
	360,000,000 Yrs	CARBON-IFEROUS	AGE of COAL BEARING PLANTS	AGE of SPORE BEARING PLANTS
	408,000,000 Yrs	DEVONIAN	AGE of FISHES and CORALS	SPORE
	438,000,000 Yrs	SILURIAN	AGE of INVERTEBRATES	SEA-WEEDS
	505,000,000 Yrs	ORDOVICIAN		
	570,000,000 Yrs	CAMBRIAN		
PROTEROZOIC	700,000,000 Yrs	PRECAMBRIAN	RISE of INVERTEBRATES	AGE
ARCHAEOZOIC	4,600,000,000 Yrs			

50

It is admitted that C-14 has credibility in determining the ages of items in the lower range of 3,000 years or so. But many thousands of things have been C-14 tested, some with very shocking results. The following information comes from several issues of a scientific journal called *Radiocarbon*. Keep in mind that things older than 50,000 years could not have enough C-14 to measure if indeed they are really that old.

According to the popular traditions of our day this chart is intended to portray the progressive "ages" of Earth's history.

OBJECT TESTED	SUPPOSED AGE (BY THE GEOLOGIC CHART)	C-14 TEST RESULT
Saber-toothed Tiger	100,000 - 1,000,000	28,000
Natural Gas	50,000,000	34,000
Coal	100,000,000	1,680

IS THERE EVER ANY AGREEMENT BETWEEN THE GEOLOGIC CHART AGES AND ACTUAL RADIOMETRIC TESTS ON THE MATERIALS FOUND AT THE VARIOUS IDENTIFIED LEVELS?

ANSWER? NO

CHALLENGE: If the geologic chart has any basis in fact, and . . .

If radiometric dating methods have any absolute scientific accuracy . . .

Why won't someone undertake a study to simply produce a complete analysis of the entire sequence?

Keep in mind that the so-called **"geologic periods" are named entirely with the theory of evolution in mind. The names are totally arbitrary; they mean nothing except what believers in the theory want them to mean.** Most of the names were assigned over a hundred years ago, long before radiometric methods were known. Consider the following:

"Paleo-zoic" means ancient life

"Meso-zoic" means middle life

"Ceno-zoic" means recent life

"Cambrian" refers to the ancient name for Whales in Great Britain.

From the *Bulletin* of the Geological Society of America (vol. 69, January 1958) Curt Teichert writes, *"NO COHERENT PICTURE OF THE HISTORY OF THE EARTH COULD BE BUILT ON THE BASIS OF RADIOACTIVE DATINGS."*[30]

Evolution demands millions of years. But if millions of years are mythical and evolution is untrue, then the whole foundation of Historical Geology vanishes in the tide that destroys every house built on error.

THE EARTH IS YOUNG!

Unlocking the Mysteries of
CREATION

Can The Rock Layers Of The Earth Be Interpreted Another Way?

In our generation, when most people focus attention on the massive layers of mountain formations, they tend to automatically think of extreme age in terms of millions of years. It is necessary to remember that such thinking has its roots in the theory of evolution. (Evolution is covered in more detail in session two of this book.)

The Familiar GEOLOGIC CHART of the history of the Earth was specifically designed with evolution in mind. Gradual processes of erosion and mountain building are supposed to have happened so slowly that only hundreds of millions of years could possibly account for it all.

WHAT FORMED EARTH'S ROCK LAYERS?

If you were to travel anywhere in the world to see the mountains and valleys that are exposed on the surface you would commonly see three things.

Sedimentary layers of sandstone, limestone, and shale are found widely. They were clearly deposited by moving water. These layers continue in every direction for many miles. Some of them are only inches thick while others are many feet thick. For the most part, our entire Earth is overlaid by up to several miles of layer after layer of sedimentary rock.

Secondly, volcanic rock layers also extend over much of the Earth. Often they are interspersed among the sedimentary layers.

Also, fossilized plants and animals are often found embedded in the earth's layers, and virtually always in mass burials. When you find fossils at all you find jumbled piles of them. Would you expect the condition of their burial to tell us much about the events responsible for burying them?

HOW DOES THE BIBLE EXPLAIN IT?

According to the Bible record, the flood was a world-wide catastrophe lasting over a year. Over 200 non-Biblical historical accounts from tribes and nations around the world also claim the reality of this cataclysm. (Seminar number two of Unlocking The Mysteries of Creation covers the subject of the flood extensively in another volume.)

With a massive upheaval such as the flood described in the Bible, we would expect to see vast sedimentation on a global scale. Volcanism and fossilization are also evidences of rapid catastrophic action.

When you realize that a single local flood, volcanic eruption, or tidal wave can powerfully wipe out vast areas and redeposit sediment dozens of feet thick in just moments, it becomes clear what enormous potential for destruction there is in a **world-wide flood.**

The rock layers, mountain upthrusts, and many erosional features of Earth are easily accounted for by the flood. The problem with many people's thinking is that they make the flood too limited when, in fact, it was the most awesome destruction our planet has ever endured.

The imaginary geologic chart with its "millions-of-years-long" epochs is totally meaningless if a world-wide flood really did happen. And after seeing the many evidences considered in this session, one fact becomes very apparent –

THE EARTH IS YOUNG!

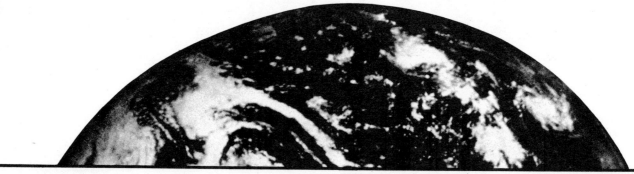

SUPPLEMENT
TO SESSION ONE

EXAMINATION OF THE "GAP THEORY" OF EARTH'S ORIGIN

"TEST ALL THINGS;

HOLD FAST WHAT IS GOOD."

1 Thessalonians 5:21

Theological Question Raised: Is there a Gap of time between Genesis 1:1 and 1:2?

What is this all about?

Many conservative Christian scholars have held to a different interpretation regarding the Biblical idea of the Earth's beginning.

This is called the "gap theory" by many. According to this theory a large period of time, possibly millions of years, may have passed between the completed creation of verse one of Genesis one and verse two. The description, "without form and void" is taken to mean that the entire creation became ruined and destroyed prior to the virtual "re-creation" described through the rest of chapter one.

The supposed destruction of the world at this time is said to have been the result of the fall of Satan. The theory maintains that the entire Earth was filled with plants and animals, and that even a race of "pre-Adamic" men ruled the Earth. Then, with Satan's rebellion, darkness invaded the perfect Earth and a global flood destroyed everything.

The vast ages of the geologic chart continued during this time gap. The fossilized plants and animals of our Earth's layers are supposed then to be the remains of that originally perfect Earth, *destroyed prior to the six literal days of re-creation* recorded in Genesis.

Where did this idea originate?

Some have said that Genesis 1:28 hints that all this may have happened. The King James Version reads that God told the first human couple to "be fruitful, multiply and replenish the Earth . . . " The thought here is that the destroyed civilization had to be replaced and the Earth **REFILLED** with inhabitants. However, the original word simply means "fill" and is thus translated in many of the up-to-date versions.

What is the history of this theory? Actually it goes back to the year 1814 and before. In Great Britain scientists were developing new ideas about the causes of Earth's geologic features. Scottish theologian, Dr. Thomas Chalmers, proposed the *"ruin- reconstruction"* ideas as a way to make the Bible record fit. The idea gained in popularity. Through the course of the 1800's evolutionary thoughts about the extreme age of the Earth and universe were highly promoted. The public became indoctrinated with the idea of the Earth being millions and billions of years old. The long-held conservative view of the Bible account was under fierce attack in academic circles.

The gap theory was further popularized with the publication of George Pember's book, *Earth's Earliest Ages,* in 1876. But the idea really gained wide acceptance after it appeared in the footnotes of the Scofield Reference Bible in 1917.

If you want to have a more thorough study on this I recommend you get a copy of John Whitcomb's book, The Early Earth.

Now, I realize that there are many good, sincere Christian scholars— men and women anointed of God—who have accepted and taught the gap theory as true. There are several Bible passages that are used to support the idea. But let's be honest enough to realize that it is *only a theory*. Let's also recognize that there are some *genuine difficulties with it.* I used to think the gap theory was a pretty good way to fit the evolutionary idea of millions of years into the Bible. After all, if you can't fit them in between Genesis 1:1 and 1:2 where else can you put them?

Analysis of Gap Theory Problems

Let's take a few minutes to analyze the problems raised by the gap theory.

1. The first big problem comes up before you even get out of Genesis one. In verse 31 God saw ALL that He made and *"behold, it was VERY GOOD."*

● What does *"very Good"* mean? Had the Earth been destroyed leaving billions of fossilized animals as evidence of the destructive judgement? Could such a world legitimately be called *"very good?"*

Was Adam created to walk over a virtual graveyard filled with all manner of life forms which God had created and subsequently destroyed? This brings up another problem.

2. When did death begin?

● According to the Bible did death on this Earth begin with some Satanically ruled race?

● Romans 5:12 plainly says: *" . . . by one MAN sin entered into the WORLD, and death by sin . . . "*

● First Corinthians 15:21 clearly relates to Adam, the first man, and the only "first" man the Bible ever acknowledges. It says, *"by a man came death."*

● Even those who believe the Gap theory admit that the first few verses of Genesis are clearly lacking a context of judgement and death.

● And keep in mind that the supposed flood of destruction following aeons of time did not occur until the fall of Satan. Doesn't the Bible show that the *"groaning and travailing in pain"* of the animal kingdom is a result of the curse brought on by Adam? Could animals have died for millions of generations even before Satan's fall?

That brings up another problem.

3. When DID Satan fall?

● We know that when Satan was created he was good and without rebellion. When was Satan created?

● In Genesis 2:1 it says: *"Thus the heavens and the Earth were completed, and all their hosts."* Exodus 20:11 elaborates further: *" . . . in six days the Lord made the heavens and the Earth, the sea and all that is in them."* Other Bible contexts refer to the all-inclusiveness of the creation event. Nehemiah 9:6 is an example. Here it says: *"Thou hast made the heavens, the heaven of heavens with all their host, the Earth and all that is on it, the seas and all that is in them."*

● Some may say that such an idea doesn't give a long enough existence for Satan before Adam's creation. I simply have to ask the question: Is such a long time required Biblically?

● After all, how long after his creation did Adam disobey God by eating of the tree of knowledge of good and evil? Probably not a very long time at all. We know that all of Adam's days totaled 930 years. His first commission was to be fruitful and multiply. Eve's first child would no doubt have been conceived within just weeks or even a few days after her creation. [The Bible calls Eve the *"mother of all"* humans, and we also realize that *"all (humans) have sinned . . . "*] If Adam fell in sin in such a short time, do you think it's possible that Satan also could have fallen in a relatively brief time after his creation?

● When God saw all He had made was it ALL very good? If Satan had already fallen by the sixth day of creation was God just ignoring him in this statement? Did God really mean *"everything I made is good except for Satan and his rebellious angels?"*

4. Who did God create to be the dominator of this Earth anyway? Does the Bible ever indicate that anyone else but Adam was specifically made to rule this Earth?

● Genesis 1:26 says: *" . . . Let us make man in our image . . . and let them rule . . . over all the Earth."*

● Were there ever animals made by God (like Dinosaurs) over which Adam did not have dominion?

Some suggest that Satan created some of the grotesque giants of the prehistoric past. But does Satan have the power to create life? Or is he the one ultimately responsible for the perversion of life? Does the Bible say anything about a demonically linked corruption of Earth's animal life? (For an in-depth study of this subject refer to Unlocking The Mysteries of Creation, Seminar number two.)

5. And what about the Great Flood of Noah's Time? The supposed flood of the gap theory left the Earth *"without form and void."* But many have assumed that the geologic history of the Earth took place before the creation of Adam and the six day creation account in Genesis. In effect then, all the major fossil-bearing rock formations were produced by a catastrophe summed up in less than a dozen words of Genesis 1 verse 2. But such an idea causes a great problem in regard to the Genesis flood of Noah's time. We have two choices:

First, was the flood of Genesis 6, 7 and 8 merely a local flood, relatively insignificant in terms of geologic formations?

Our second choice poses a real dilemma. If the flood of Noah was an Earth-shattering year-long catastrophe that totally rearranged the structure of Earth's crust, then evidence of a previous destruction would have been obliterated. (For more information on the World-Wide Flood: Myth or Proven Fact, refer to Seminar number two of Unlocking the Mysteries of Creation.)

All this discussion really brings us back to where we started: When did it all begin?

If the Earth really is only a few thousand years old and the present land forms were produced by the awesome destruction of the Noahic flood, then we don't really need a gap, do we?

The opening verses of Genesis describe, not a destroyed Earth, but merely an incomplete Earth. The evening and the morning marked the first day in God's masterful building project.

Let's always remember that God's word is not some deep dark mystery. But as we already learned in Proverbs 8:8 and 9, God's words are straightforward. They give understanding even to the simple as it says in Psalm 119:130.

Indeed, the Bible is the key to unlock the mysteries of Creation. Be careful not to program your computer with unproven theories as if they were absolutely true.

Remember, God is not afraid of the facts. Every idea bears inspection. That's why God dares to say in First Thessalonians 5:21, *"Prove all things, hold fast to that which is good."*

The same thing applies to our ideas about salvation and our eternal existence. Be careful not to take chances about the things of God and your spirit. What you believe is the foundation upon which you build your actions. The words of Jesus have withstood the test of time. How and why is it that faith in Him alone can guarantee a restoration of relationship between you and God?

His Word tells us in Colossians chapter one that it's because of Jesus that we can be transferred out of the kingdom of darkness and into the kingdom of light. How can Jesus have so much authority? Verse 16 tells us it's because all things were created through Him. In fact the next verse says that in Jesus everything holds together. The secrets of the universe are not ever going to be found in science alone. No mere man or guru has all the answers. But think of what the Creator has demonstrated to us through Jesus Christ. And His Spirit is declaring that reality to us now.

References For Section One

1. Patten, Donald, The Biblical Flood And The Ice Epoch, Pacific Meridian Pub., Seattle, 1966, pages 194-224.
2. Whitley, D.G., "The Ivory Islands in the Artic Ocean," Journal of the Philosophical Society of Great Britain, XII (1910), p. 49 in Earth In Upheaval, by Immanuel Velikovsky, Dell edition, 1955, p. 19.
3. Bible-Science Newsletter, January 1975, p. 6.
4. Morris, Henry, Scientific Creationism, Creation-Life Publishers, San Diego, 1974, p. 152.
5. Ibid., p. 156.
6. Slusher, Harold, "Some Astronomical Evidences for a Youthful Solar System," Creation Research Society Quarterly, v.8, 6/71, p. 55-57.
7. Cook, Melvin, Prehistory And Earth Models, Max Parrish, London, 1966.
8. Morris, Ref. 4, p. 155.
9. Blick, Edward, Correlation Of The Bible And Science, Southwest Radio Church, Oklahoma City, 1976, p. 28.
10. Allen, Benjamin, "The Geologic Age Of The Mississippi River, Creation Research Society Quarterly, v.9, 9/72, p. 96-114.
11. Patten, Ref. 1, p. 11.
12. Whitcomb, John, and Morris, Henry, The Genesis Flood, The Presbyterian And Reformed Publishing Co., Philadelphia, 1961, p. 408-409.
13. Morris, Ref. 4, p. 156.
14. Whitcomb, John, C., Jr., The World That Perished, Baker Book House, Grand Rapids, 1973, p. 114.
15. Whitcomb and Morris, Ref. 12, p. 392-393.
16. Morris, Ref. 4, p. 167-169.
17. Barnes, Thomas G., Origin And Destiny Of The Earth's Magnetic Field, Institute for Creation Research, San Diego, 1973, in Ref. 4, p. 157.
18. Morris, Ref. 4, p. 153.
19. Cook, Melvin, "Where is The Earth's Radiogenic Helium?" Nature, v. 179, 1/26/57, p. 213, in ref. 9 (above), p. 26.
20. Origins Film Series Handbook, Films For Christ, 1983, p. 30.
21. Akridge, Russel, "The Sun Is Shrinking," Acts and Facts Impact No. 82, Institute for Creation Research, San Diego, 4/80.
22. Slusher, Harold, article in Bible-Science Newsletter, January 1975, p. 2.
23. Ibid.
24. Ibid.
25. Whitcomb, John C. Jr., The Origin Of The Solar System, Presbyterian And Reformed Publishing Co., Philadelphia, 1976.
26. Morris, Ref. 4, p. 147.
27. McLean, Glen S., personal interview in 1984.
28. Whitelaw, Robert, "Time, Life, and History in the Light of 15,000 Radiocarbon Dates," in Speak To The Earth, edited by George Howe, Presbyterian and Reformed Publishing Co., 1975, p. 339.
29. Kofahl, Robert E., Handy Dandy Evolution Refuter, Beta Books, San Diego, 1977, p. 119 citing several scientific journals.
30. Whitcomb and Morris, Ref. 12, p. 365.

Selected Reading

Ackerman, Paul D., It's A Young World After All, Baker, 1986.

Baugh, Carl E., Panorama Of Creation, Southwest Radio Church, Oklahoma City, 1989.

Dillow, Joseph, The Waters Above, Moody Press, Chicago, 1982.

Ham, Ken, The Genesis Solution, CLP, 1988.

Morris, Henry, Scientific Creationism, CLP, 1974.

Morris, Henry, and Parker, Gary, What is Creation Science?, Master Books, 1982

Whitcomb, John C. Jr., The Early Earth, Baker Book House, Grand Rapids, 1972.

Whitcomb, John C., and Morris, Henry M., The Genesis Flood, The Presbyterian and Reformed Publishing Co., Philadelphia, 1961.

Rehwinkel, Alferd M., The Wonders Of Creation, Baker Book House, Grand Rapids, 1974.

Seagraves, Kelly, Jesus Christ, Creator, Creation Science Research Center, San Diego, 1973.

SESSION 2

UNLOCKING THE MYSTERIES OF EVOLUTION

*"ANY STORY SOUNDS TRUE
UNTIL SOMEONE TELLS THE OTHER
SIDE
AND SETS THE RECORD STRAIGHT."*
Proverbs 18:17 (TLB)

FIRST OF ALL, YOU MUST UNDERSTAND THAT IN THE LAST DAYS SCOFFERS WILL COME, SCOFFING AND FOLLOWING THEIR OWN EVIL DESIRES. THEY WILL SAY, "WHERE IS THIS 'COMING' HE PROMISED? EVER SINCE OUR FATHERS DIED, EVERYTHING GOES ON AS IT HAS SINCE THE BEGINNING OF CREATION." BUT THEY DELIBERATELY FORGET THAT LONG AGO BY GOD'S WORD THE HEAVENS EXISTED AND THE EARTH WAS FORMED OUT OF WATER AND BY WATER. BY THESE WATERS ALSO THE WORLD OF THAT TIME WAS DELUGED AND DESTROYED. BY THE SAME WORD THE PRESENT HEAVENS AND EARTH ARE RESERVED FOR FIRE, BEING KEPT FOR THE DAY OF JUDGMENT AND DESTRUCTION OF UNGODLY MEN.

2 Peter 3:3-7

Contents Of Section Two

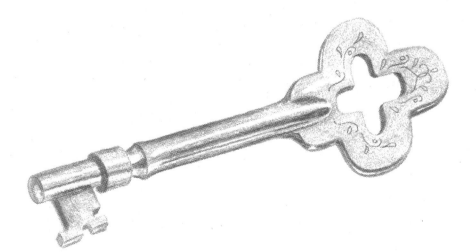

ORIGINS: BY DR. WERNHER VON BRAUN

"One cannot be exposed to the law and order of the universe without concluding that there must be design and purpose behind it all . . . The better we understand the intricacies of the universe and all it harbors, the more reason we have found to marvel at the inherent design upon which it is based . . . To be forced to believe only one conclusion—that everything in the universe happened by chance—would violate the very objectivity of science itself . . . What random process could produce the brains of a man or the system of the human eye? . . . They (evolutionists) challenge science to prove the existence of God. But must we really light a candle to see the sun? . . . They say they cannot visualize a Designer. Well, can a physicist visualize an electron? . . . What strange rationale makes some physicists accept the inconceivable electron as real while refusing to accept the reality of a Designer on the ground that they cannot conceive Him? . . . It is in scientific honesty that I endorse the presentation of alternative theories for the origin of the universe, life and man in the science classroom. It would be an error to overlook the possibility that the universe was planned rather than happening by chance."

[Dr. von Braun was a leading scientist in the U.S. space program until his death in the late 1970's. These remarks from an article in Applied Christianity were published in the Bible Science Newsletter, May 1974, p. 8]

Evolution's Roots

"Know this first of all, that in the last days mockers will come with their mocking, following after their own lusts, and saying, 'where is the promise of His coming? For ever since the fathers fell asleep, all continues just as it was from the beginning of creation.'"

2 Peter 3:4

Until the mid-1800's brilliant scientists had no trouble accepting the Biblical account of creation. Remember Isaac Newton? Other familiar ones include Kepler, Faraday, Kelvin, Mendell, Lister, and Pasteur. What happened? They all died with faith in God, but a new generation arose in the midst of much skepticism and moral decline in society.

UNIFORMITARIANISM

Charles Lyell, one of the pillars of modern geology, published his *Principles of Geology* in the early 1800's. It has literally shaken the intellectual world. The study of Earth's geophysical structures was an infant science then. But now his concepts are so widely accepted that most who believe them don't even know who originated them.

Lyell's basic premise has been adopted by all scientific fields even though his field of study was Earth science. He proposed that all our Earth's features can be explained in terms of present observable processes, which supposedly have always gone on in the past. Thus:

"THE PRESENT IS THE KEY TO THE PAST"

From his system of reasoning a doctrine of origins was devised by simply projecting all natural processes back through time indefinitely. Catastrophes like the flood were strictly ruled out.

CHARLES DARWIN

In our time the word "evolution" and "Darwin" are almost inseparable. At the age of 22 in 1831, with a basic degree in theology, Charles Darwin began a 5-year world voyage as the naturalist aboard the Ship Beagle. He read Lyell's book and began gearing his concepts of biological origins according to the uniformitarian philosophy.

Darwin's dogmas of the evolution of species could never have gotten off the ground without Lyell's "gift" of billions of years. Thus, the foundation was laid for what came to be know as Darwinian evolution.

Though the theory of evolution, and even the idea of natural selection, were not original to Darwin, it was his book, *The Origin Of Species* (published in 1859), which quickly exalted him to fame. He became the central figure of evolutionary thinking in a world filled with revolution, humanist attitudes, and spiraling social upheaval.

GRADUALISM

The word gradualism is often used in place of uniformitarianism since the intended meaning is the same. Slow processes gradually going on through immense spans of time are assumed to be the only cause for virtually everything in nature and in the universe.

A WORKING DEFINITION OF EVOLUTION

"Evolution . . . a directional and essentially irreversible process occurring in time, which in its course gives rise to an increase of variety and an increasingly high level of organization in its products. Our present knowledge indeed forces us to the view that the whole of reality is evolution, a single process of self-transformation." Julian Huxley

THE EVOLUTIONARY TREE OF LIFE

Most school children today are familiar with the typical diagram depicting how all forms of life supposedly evolved through the last two thousand million years. This chart is usually published with no explanation that it is purely a theory with no evidence to substantiate it.

IF THERE IS NO EVIDENCE, WHY BELIEVE THE THEORY?

Many scientists have faced the reality that the subject of origins comes down to a personal choice of opinion. The statement of S.A. Keith may help clarify the issue: "Evolution is unproved and unprovable. We believe it because it is the only alternative to special creation, and that is unthinkable." [1]

WHAT IS SCIENCE?

The Oxford Dictionary defines science as "a branch of study . . . concerned either with a body of demonstrated truths or with observed facts systematically classified . . . under general laws, and which includes trustworthy methods for the discovery of new truth within its own domain."

G.G. Simpson, a leading evolutionist, has said: "It is inherent in any definition of science that statements that cannot be checked by observation are not really about anything . . . or at the very least, they are not science."

TRUE SCIENCE IS
- **Observable**
  - **Demonstratable**
    - **Repeatable**

DOES SCIENCE CONTRADICT THE BIBLE?

Some feel that God's word and His world are separate and possibly conflicting in apparent information. However, keep in mind that the word of God speaks unapologetically about the origin and operation of the natural world. The concept of creation was approved by Christ and the apostles who wrote the New Testament.

SCIENCE MYSTERIES

Despite all the work that has been done in all the science disciplines there are major enduring problems. What is electricity? What is light? What is gravity? Why do these phenomena behave so consistently? Some features of these things are known but not the basic "why" behind them.

CAUTIONS FOR SCIENCE

- The many realms of science make being an authority in more than one realm impossible for most men.
- Scientists are real people; they have the same human weaknesses and biases as non-scientists.

- Scientists who wish to be credible must be cautious not to make broad declarations on issues of religion or human spiritual matters in the name of science.

- When scientists do say things about origins, supernatural powers and spiritual things, you should be careful to examine and challenge them. Theories are not facts!

WHAT IS A THEORY?

A theory is a guess or suggestion that spawns inventive and logical research to explain some natural phenomenon.

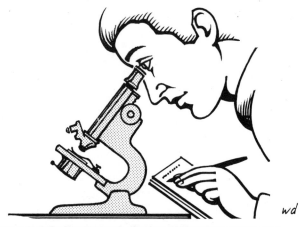

HAVE SCIENTISTS BEEN KNOWN TO BE WRONG?

False ideas in the name of science have been accepted in past generations only to be discarded by their descendants.

- The flat Earth idea

- The idea that bloodletting is appropriate medical procedure for reducing fever

- The geocentric theory - the idea that the Earth is the physical center of the universe.

- Phlogiston was believed by 17th and 18th century scientists to be the magical ingredient of all combustible things which enabled them to burn. French chemist, Lavoisier, showed that oxygen was the key for fire (rapid oxidation) to burn.

IS EVOLUTION SCIENTIFIC?

No matter how you look at it, the theory of evolution must trace back to a point where inanimate matter became a living form. Look at the foundation of this theory.

• **UNKNOWN CHEMICALS** in the primordial past ... **through** ...

• **UNKNOWN PROCESSES** which no longer exist ... produced ...

• **UNKNOWN LIFE FORMS** which are not to be found ... but could, through

• **UNKNOWN REPRODUCTION METHODS** spawn new life ... in an ...

• **UNKNOWN ATMOSPHERIC COMPOSITION** ... in an ...

• **UNKNOWN OCEANIC SOUP COMPLEX** ... at an ...

• **UNKNOWN TIME** and **PLACE.**

Is it any wonder why many scientists like Dr. Henry Morris (who proposed the above idea) insist that evolution does not even constitute a bona fide scientific theory!

THE BIG QUESTION

Bible believing Christians are often mocked by those who say:

"Prove your eternal God created all this!"

But what's the alternative? Why not ask the scorner:

"Prove to me that atoms are eternal! If God is not the Creator, how did the atoms get there?"

MYSTERIES OF EVOLUTION?

For any Creation or upward development there are basically three things which are absolutely essential:

1. A PROGRAM to direct the growth.

2. A MECHANISM to energize the growth.

3. A PROTECTION SYSTEM to sustain the growth.

DOES CHANGE EQUAL EVOLUTION?

Everyone who cares to investigate can see that change is constantly occurring in natural systems. Some call this "micro-evolution." But let's realize that change is always limited within "kinds." The commonly held belief in Evolution can be more specifically called "macro-evolution." It requires more than minuscule mutations to produce all the different living kinds. Indeed, TRANSMUTATION is the order of the philosophy.

"Catch the foxes for us, the little foxes that are ruining the vineyards while our vineyards are in blossom."

The Song of Solomon 2:15

FIVE
FUNDAMENTAL
FALLACIES

FIVE FUNDAMENTAL FALLACIES

#1 It All Started With A Big Bang

WHAT STARTED IT ALL?

According to the evolutionary philosophy, all life ultimately originated from a super explosion!

Currently, many believe that this mega-blast happened some 12 to 15 billion years ago. An infinitely tiny mass of matter somehow chanced to blow up and things have been spinning off that explosion ever since.

From then on, with random processes, everything in the universe became increasingly complex. The sun and all the other stars eventually condensed out of gas. The planets, including our Earth, also condensed out of dust particles swirling through space.

BUT WAIT! We need to ask an important question here.

DO EXPLOSIONS EVER INCREASE ORDER?

TEST: If you take a stick of dynamite and go out and blow up a pile of lumber or other building material, is it going to end up more orderly?

BUT OTHERS ARGUE: "Given a steady addition of energy it's possible for order to come about. After all, the Earth is an open system to the sun's immense input of energy."

HOLD IT! Another question must be asked here:

DOES ENERGY ALONE PRODUCE ORDER?

Try another one:

HOW MANY LIGHTNING BOLTS DOES IT TAKE TO BRING A CORPSE TO LIFE?

Remember Frankenstein? Miracles like that only happen in the movies!

WHAT DO EXPLOSIONS REALLY CREATE?

When Mount St. Helens blew up in the Northwest U.S.A. we saw a good demonstration of the results of an explosion. It was a tremendously powerful blast, equal to many times the force of a nuclear bomb explosion. For miles around the mountain everything was desolated. In fact it was called a "dead zone."

Scientific articles repeatedly refer to the "big bang" as a reasonable explanation for the origin of the universe. But there are physical realities in the heavens which defy the possibility of such an explosion being responsible for the order we see.

If the big bang were true:

1. The planets should all rotate in the same direction on their axes. But Venus and Uranus both rotate backwards.

2. All the nearly 50 moons of the solar system should orbit their respective planets in the same direction. But at least 11 of them orbit in the opposite direction.

3. The orbits of the moons of the several planets would be expected to lie flat over the equatorial planes. However, a number of moons, including Earth's, follow a highly tilted orbit.

4. The sun's inner planets should be made up mostly of hydrogen and helium, like the sun. Much less than 1% of the Earth is made of these elements.

There are a number of other "mysteries" to be found in the real universe that simply do not add up to a big bang for its origin.

ASTRONOMICAL "FINE TUNING?"

In the June 1983 issue of National Geographic (page 741) one astronomer is quoted concerning the remarkably puzzling precision of the universe:

"To get a universe that has expanded as long as ours has (an assumption) without collapsing or having its matter coast away would have required *extraordinary fine tuning*. A Chicago physicist calculated the odds of achieving that kind of precise expansion . . . would be the same as throwing an imaginary microscopic dart across the universe to the most distant quasar and hitting a bull's eye one millimeter in diameter."

THINK! Knowing the odds, why don't more astronomers confess that presently popular theories are totally inadequate to explain such astronomical precision?

Unlocking the Mysteries of
CREATION

FIVE FUNDAMENTAL FALLACIES

#2 Non-Living Matter Produced Life

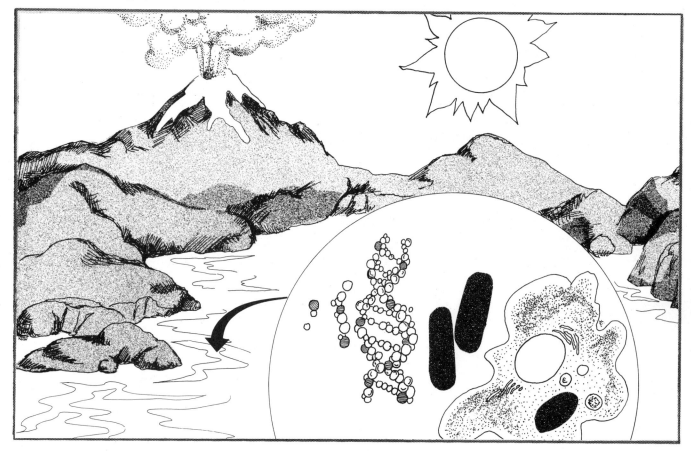

SPONTANEOUS GENERATION

With random processes and time, the raw elements of the Earth eventually became living organisms. This widely held belief is called spontaneous generation. Have you seen the pictures used to try to impress you with what some feel represents reality?

A COMMON DESCRIPTION OF HOW THE THINKING GOES

The following quote is taken from a book by Benjamin Bova called *The Giants Of The Animal World.* Carefully notice the wording and the implications of this statement. It's quite typical of explanations given about how the first life came to be.

"**The Earth was born about 5 billion years ago . . . Scientists are *not certain* of just how it happened, but they *believe* that life began about 2 billion years ago in the shallow waters of some *unknown* seacoast. A group of atoms came together in a *very certain way*. They *formed* a new type of molecule—a giant molecule, much bigger than all the other atom groups nearby. This large molecule *could do something* that no other molecule could do. It was able to take simple atoms and smaller molecules from the sea and make a new giant molecule just like itself.**"

DARE TO PROVE ALL THINGS

(1 Thessalonians 5:21)

Note: (from the above statement)
   "scientists are not certain"
   "but they believe"
   "a very certain way"
   "They formed a new type"
   "molecule could do something"

MEDIEVAL SCIENCE REVISITED

In the middle ages virtually everybody in Europe believed that **rats and flies came from garbage.** After all, every town had a garbage dump on the outskirts and rats and flies were always seen emerging there. **It was an observed scientific fact!**

Finally an inquisitive researcher named Francesco Redi decided to test the accepted theory. In 1668 he isolated some garbage and just watched. What do you suppose happened?

You guessed it. Nothing!

By the mid-1800's the famous scientist, Louis Pasteur, confirmed this reality that seems so obvious to us today.

So what's the conclusion? Imagine! Rats and flies can only come from parent rats and flies! A new scientific law was discovered. It's called the law of:

BIOGENESIS

Bio- means life. Genesis means beginning. Thus biogenesis has come to be known as one of the most universal laws of science: life begins from life! It always works that way! Consequently, the theory of spontaneous generation was discarded. It was obsolete and scientifically unacceptable. Unacceptable, that is, until the modern "scientific" illusion of evolution came along. **And evolution doesn't even have the garbage!** Evolution requires that life came from the raw elements of the sea, and nothing more.

What does the Genesis record say?

"God said: 'Let the Earth bring forth living creatures after their kind: cattle, and creeping things, and beasts of the Earth after their kind;' and it was so." Genesis 1:24.

God is the one supremely powerful source of living creatures. All of them were created to reproduce after their kind.

#3 Time: The Magic Factor

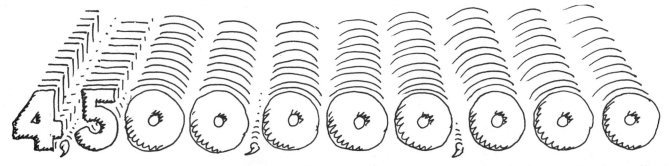

Evolutionary ideas require lots of time. Does 4 and 1/2 billion years sound sufficient? In fact evolutionary speculation seems to keep raising the age of the earth at about the rate of doubling every 30 years or less! Recently they've been insisting on four point SIX billion years, but "given enough TIME" that may be soon protracted to a full five billion.

Suppose you were a high school student assigned to write a term paper for your biology teacher. Imagine the look on his face if you proposed to elucidate on the theory that modern man evolved from single-celled amoeba in just 2 seconds. "Ridiculous," he would say, and rightfully so. But you are determined to come up with something new and scientifically stimulating so you think awhile and return with a new twist. "How about the idea that man evolved from amoeba in two billion years?" you ask. With a wry smile on his face, your teacher smugly tells you, "Now you've got the right idea."

Do you see what happened? Simply give it billions of years and:

THE RIDICULOUS BECOMES ACCEPTABLE

The story in which the kiss of a princess turns a frog to a prince is called a what? A FAIRYTALE! Now we see people waving the magic words of "MILLIONS OF YEARS" and it's called "SCIENCE"... evolutionary "science" that is.

BUT IS IT SCIENCE?

WHAT REALLY DOES HAPPEN AS TIME MARCHES AHEAD?

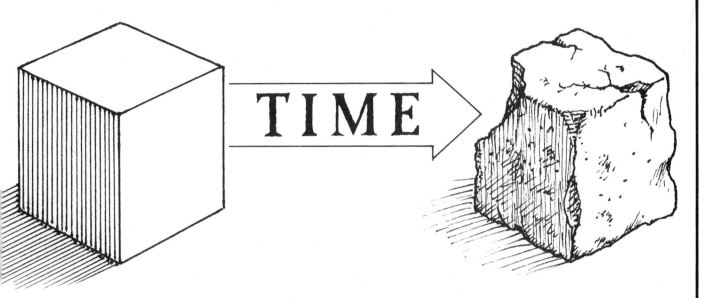

- Orderly things become disordered.
- New things get old and break down.
- Living things age and wear out.

This is called the second law of thermodynamics and it is one of the most constant laws of nature, found everywhere in the universe.

It is sometimes referred to as the law of ENTROPY.

WHICH REPRESENTS REALITY?

The Latin word from which "evolution" comes literally means "an out-rolling." It indicates something spiraling from the infinitesimal to all of reality. The word "entropy" comes from the Greek word meaning "in-turning."

Notice the polarity of difference:

Evolution - change outward and upward

Entropy - change inward and downward

Every energy system in the universe wears out. Stars burn out, galaxies fly apart, energy becomes less and less useful through time.

Science is painfully aware of the rigidity of this natural law. The impossibility of the perpetual motion machine verifies the consistent observation that everything in our physical universe is running down.

Unlocking the Mysteries of **CREATION**

THE TWO LAWS EXPLAINED

FIRST LAW: Energy is conserved.

All existing processes of nature merely change energy from one form to another. In nature, energy is neither created or destroyed. Matter itself (which is potential atomic energy) is maintained at a constant level. Processes change matter and energy from one form to others but the total quantity of energy in the universe always remains the same.

SECOND LAW: Energy dissipates.

As processes in nature occur, the total energy reservoir is reduced to simpler forms with a consequent increase in what has been termed "entropy." As energy is used it becomes less available for further use. Part of the energy spent to produce something is always lost by radiation, friction, or other effects. It becomes non-recoverable heat dissipating in space. Ultimately, as things are going, the entire universe will end up being filled with a stagnant mass of low-level heat energy.

THERMODYNAMICS AND SCRIPTURE

No discovery of true science takes God by surprise. We should expect that, in His wisdom, there would be hints of these truths found in His word, the Bible.

1. The first law speaks of a total creation, originally completed, and now sustained by God's power.

2. The second law speaks of the curse of decay and death, brought on by man's sin, and causing an overall degeneration in everything.

THE FIRST LAW IN SCRIPTURE

The "conservation principle" of the first law of thermodynamics is easily understood in the context of a completed creation which is now being sustained by the Creator.

"...All things were created through Him.." (past tense; notice it is not a continuing process).

"...in Him all things consist." (Greek word for sustain; i.e. nothing is lost from it).
Colossians 1:16 & 17

"...He made...upholding all things by the word of His power."
Hebrews 1:2 & 3

". . . by the word of God . . . heavens and Earth . . . reserved . . . kept in store . . ."
2 Peter 3:5 & 7

". . . He commanded . . . created . . . established them forever."
Psalm 148:5 & 6

". . . He . . . created . . . not one faileth."
Isaiah 40:26

". . . Lord . . . made . . . all things . . . preservest them all."
Nehemiah 9:6

It is clear from Genesis chapter one and the first three verses of chapter two that whatever methods God used to create were stopped then. *"He rested from all His work"* and called it *"very good."* (Genesis 2:2 & 1:31)

THE SECOND LAW IN SCRIPTURE

The "decay principle" is consistent with the entire scripture in light of the curse in Eden.

Psalm 102:26

"Even they will perish (the starry heavens), but Thou dost endure; and all of them will wear out like a garment . . ."

As Romans 8:22 says: *"the whole creation groans and suffers . . ."* This seems contrary to God's original purposes for a creation which He called *"very good."* The death and decay was obviously not intended to prevail as it does now. Nor will it dominate the universe in the eternal kingdom as described in Revelation 22:3.

The result of man's sin described in Genesis 3:17 has even affected the ground, the very elements of the physical creation, including the dust of which Adam was made. The second law began then. Before that, some higher law of perfection and preservation must have

been in effect. Such supernatural care was indeed in effect at least partially during Israel's 40 year episode in the wilderness. (see Nehemiah 9:21)

The Bible makes clear the fact that the second law is not permanent. The Creator Himself has promised that the curse of decay is not endless.

". . . creation . . . will be set free from . . . bondage of decay." (RSV)
Romans 8:21

". . . no more curse . . ."
Revelation 22:2

THINK!

• The second law proves there had to be a beginning once or else all creation would be dead by now.

• The first law proves that all that exists must have been created by a Designer, because no process in nature creates anything.

Unlocking the Mysteries of **CREATION**

FIVE FUNDAMENTAL FALLACIES

#4 Random Chances Result In All Complexity Of Living Things

One of the most commonly heard beliefs of evolution dogma is the idea that everything in reality is only the result of pure chance. The infinite evidence in nature of orderliness, complexity and design, is only apparent to the evolutionist and not real. The implication of such thinking is that there really isn't any order in anything because everything is just the result of random (accidental) **CHANCE!**

> *THINK!* Is it possible for the precise organization of parts in living systems to come into existence by mere chance?

WHAT IS RANDOM CHANCE?

When flipping a coin you have a chance of one in two that it will turn up heads. How likely is it that three objects would be arranged in a certain way if they were spilled out on the table? The chances can be figured by a simple mathematical equation. In this case it is expressed as "3!" and is called "three factorial." The calculation is figured this way:

$$1 \times 2 \times 3 = 6$$

Thus, there are six ways to arrange three parts in a simple sequence. When you start adding parts, the chances grow very slim quickly. For example, what are the chances of blindly arranging (by chance) 6 different items in one perfect order? The result is one out of 6! which is calculated as:

$$1 \times 2 \times 3 \times 4 \times 5 \times 6 = 720$$

For the time being, forget the problem of how the first living cell happened by chance, with its multiplied thousands of specially organized parts. Let's just look at the odds of arranging a system of 200 parts. The human skeleton has about 200 bones. What are the chances of getting those bones all put in the right order from the beginning?

So how many possible ways are there to arrange a system with 200 parts? And what are the chances of having all those parts fall into place at any given moment?

The number 200! (200 factorial) is immense. It is represented by the number 10 with 375 zeros after it. There is no way to imagine that number! Each time you try a new arrangement of your 200 parts you use up one of your "chances." Let's say you can sort out your 200 parts a new way every second. Now remember there is only one correct way out of 200! different possibilities.

How many seconds are there in a year?

60 sec. x 60 min. = 3600 sec./hr.

x 24 hours = 86,400 sec./day

x 365 days = 31,536,000 sec./year

In a billion years?

31,536,000,000,000,000 seconds

In ten billion years?

315,360,000,000,000,000 seconds

How much is that?

Under a third of a billion "billions."

But 200! amounts to 10 with 375 zeros after it. The total number of electrons that could be packed into the entire universe would only total 10 with 130 zeros after it. Do you see how utterly ridiculous this gets?

But this is just one system of only 200 parts. What about the chances of all the millions of parts of the millions of organisms on Earth coming together in certain ways with no design, no pattern, no system behind it?

Albert Einstein has been quoted as saying: "God Almighty does not throw dice." Is it any wonder why mathematicians can have serious doubts about evolution?

THE COMPLEX DESIGN OF THE SINGLE CELL

When considering the subject of chance as it relates to evolution, consider one of the most complicated single things we know about: the living cell.

Think of it! Each one of us came from one single fertilized cell. In the nucleus of that little dot the master computer of what scientists call DNA contained the genetic programming for every aspect of the yet-undeveloped adult individual. Every organ, every nerve, every hair, even personality traits and behavioral patterns as well as hair and skin color are programmed in those incredibly tiny chromosomes.

HOW TINY ARE THESE MINI-COMPUTERS?

According to Ashley Montague in his book *Human Heredity,* the space occupied by all this data is incredibly small. If you could gather the programmed genetic coding for every human being on Earth from his father's side alone it would only take up the space of less than half the size of an aspirin tablet! As impressive as micro-chips are to this comput-er generation of ours, this micro- circuitry is beyond the incredible. But somehow, random chances had to be responsible if evolution is true.

It's been estimated that a single gene is between 4 and 50 millionths of an inch across. A half million genes will easily slide around in a hole made by an ordinary pin point!

APPRECIATING GOD'S AMAZING DESIGN OF GENES

Within the nucleus of every cell twenty three pairs of chromosomes are intertwined like a wad of spaghetti in the nucleus. These are made of DNA. The DNA molecules contain all the thousands of genes which program living things to go on as they do. Genes mastermind the complexities of life in ways infinitely beyond man's understanding.

WHAT DO GENES DO?

1. Determine all inherited traits like height and personality type.

2. Direct all growth processes like when and how baby teeth are pushed out by adult teeth.

3. Program all structural details for every organ and system detail in the organism.

THE REAL SURPRISE

We now realize that each and every cell in your entire body contains **ALL THE GENETIC CODING FOR ALL THE OTHER CELLS OF YOUR BODY.**

Think of it! And somehow, despite the vast sea of "blueprints" a liver cell always "knows" its own job and keeps it. A liver cell always remains a liver cell and does not become a heart cell or something else.

The Complexity Of The Single Cell

Not so long ago the single living cell was called the "simple cell." But not any more! The bulk of the cell body was thought to be a jelly-like mass of what was called "protoplasm" which simply means "living substance." In 1963 Dr. George Palade of the Rockefeller Institute in New York discovered there is more to it than meets the eye. What he found was an amazingly intricate system throughout the protoplasm. Now it's called the "endoplasmic reticulum" (E.R. for short). This vast labyrinth of incredibly fine tubes and chains of minute bags totally permeates the entire cell body. The E.R. has been described as one of the most complicated and beautiful structures in the universe. Another long-held concept of modern science has crumbled. The idea that molecules just bang around haphazardly in the jelly-like protoplasm has been discarded because of new light on the subject.

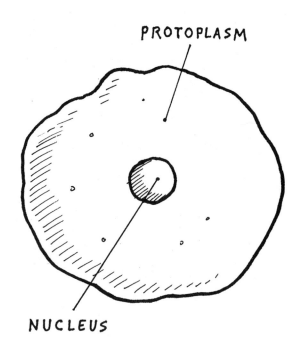

PROTOPLASM

NUCLEUS

Scientists today tell us that the single cell is more complex than a large city. As in a city, so it is in a cell. Systems are working and jobs are being done by the thousands. But unlike cities, the cell functions perfectly with no breakdowns!

The adult human may have as many as 60 trillion of these walled cities. Think of the details that are going on constantly:

● *structural design*

● *energy generators*

● *invasion guards*

● *transport systems*

● *food factories*

● *protective barriers*

● *waste disposal systems*

● *communication links within and outside the cell city*

SOME AMAZING INSIGHTS

THINK! How important is the "skin" of a living cell?

The surface membrane of a cell is amazingly delicate but equally as "powerful" in its control of the cell system. It's less than a third of a millionth of an inch thick. It controls the entry and exit of everything for the cell. It behaves almost as if it had the chemical senses of taste and smell. When a desirable molecule floats by, the membrane forms a little "finger" that reaches out and pulls the needed nutrient inside. Crucial chemical enzymes coat the skin of the cell, transferring information to and from other cells. The cell could not survive without these enzymes functioning precisely as they always do.

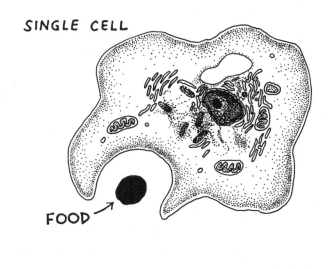

SINGLE CELL

FOOD

The RNA is the messenger in the metropolis of a cell. It looks like DNA but it has a passport to leave the nucleus. With incredible speed the RNA molecule acts like a computer printer.

Let's see what happens.

1st — The master DNA and the messenger RNA intertwine in a split second.

2nd— The DNA instantly imprints a section of its code on the RNA and then separates from it.

3rd — The RNA rushes to the edge of the cell city to transfer its code to enzymes one after another, in rapid-fire succession.

4th — Each enzyme, by this code, is commissioned to do a particular job somewhere in the larger organism.

The whole mass of cells within a body communicates by RNA. According to experts, they **"somehow cooperate"** to act like a dog, a fish, a man, or whatever the organism is supposed to be. That "somehow" is the insoluble mystery to secular scientists who choose not to begin their research on the factually solid foundation of God and His word.

"O the depth of the riches both of the wisdom and knowledge of God! How unsearchable are His judgments and His ways past finding out!" Romans 11:33

A CURIOUS NOTE

Scientists have discovered that DNA is found in the nucleus of ALL living cells with the exception of red blood cells and a few certain viruses. Is it not strange that the one component which science has singled out as the *"mysterious basis of ALL life"* (DNA) is NOT found in what God's word tells us is where the very essence of life is?

"The life of the flesh is in the blood . . ."

Leviticus 17:11

"How precious to me are Your thoughts, O God! How vast is the sum of them!"

Psalm 139:17 (NIV)

More Mysteries Of Chance

YOU'VE HEARD THE TERM BIONICS?

Bionics is the modern innovation of fabricating man-made systems parallel to designs found in nature.

Keep in mind that evolution requires that blind chance alone is responsible for all design in living creatures.

RADAR IS A BIONIC INVENTION!

THINK! Where did the idea of radar originate? Who had it first?

The curious little bat had a form of radar first! Nobody would dare say that modern radar technology just magically "happened by chance." Yet how is it that so many so-called intellectuals insist that no Designer planned the intricate radar system of the bat? Could only random accidents, which have never been witnessed by anyone, bring about all that complexity?"

The sonar of the dolphin is said to be a thousand times more sophisticated than the sonar technology of our modern nuclear submarines.

What is the bionic equivalent to the human system of remembered sight? The video camera and recorder, genius invention that it is,

can only be called "crude" when compared to the amazing complexity and sensitivity of the "living technicolor visual system," the human eye and computer brain.

IT ALL HAPPENED BY CHANCE!!

LOGIC

EVOLUTIONIST

T. EVANS

DESIGNED THINGS HAD A DESIGNER!

The very idea of all this magnificent order happening by chance with no Designer has about as much chance as a monkey sitting down at a typewriter and pounding out the entire dictionary letterperfect. Yet, as absurd as that is, evolutionists insist it must have happened that way: by chance!

THINK!

Let's say you were out taking a stroll one day along the lakeshore. You happened to look down and spot a piece of flint that catches your attention. You bend down to pick it up ... brush off the sand ... and ... low and behold, you realize you've discovered an ancient Indian arrowhead. You marvel at its careful design; a simple bit of technology yet profoundly informative. What does it tell you?

SOMEONE MADE THIS! You realize the fact instinctively. There is no doubt about it. The symmetry of the little point gives it away. It is not the product of random erosion over the years as the waves of the lake lapped its shores. No! A thinking person designed and formed this little tool.

Now, turn to the wild flower just a step away. There's a honeybee hovering over its bright petals. You ponder the design, the complexity, the detailed intricacy of that little plant-animal relationship. What a wonder you are experiencing.

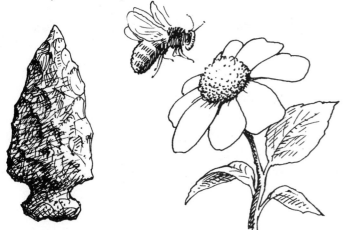

"WHO KNOWETH NOT IN ALL THESE THAT THE HAND OF THE LORD HATH WROUGHT THIS?"

Job 12:9 (KJV)

The idea of computers, or any other relic of technology, just happening by sheer chance, (let alone the astounding complexity of the human brain) has about as much chance as a monkey typing the unabridged dictionary perfectly by chance (i.e. NONE).

But wait! Hasn't life been "created" synthetically in a test tube?

JUST THINK, AS SOON AS I PRODUCE LIFE IN THE TEST TUBE I WILL DISPROVE THAT ABSURD IDEA THAT AN INTELLIGENT BEING WAS NEEDED TO DO IT IN THE BEGINNING!

T. EVANS

NO!

What has been done?

Amino acids, building blocks of proteins (which, in turn, are a part of all living cells) have been experimentally produced in controlled laboratory tests. This is a far cry from the production of "life."

THINK! Enzymes can only be produced by living cells, yet living cells absolutely require many specialized enzymes in order to survive!

Is it any wonder why some leading evolutionists have finally faced reality? Sir Fred Hoyle is one. He says the chances of life evolving without intelligent design are comparable to a tornado ripping through a junk yard and producing a Boeing 747 all by itself!

Unlocking the Mysteries of
CREATION

FIVE FUNDAMENTAL FALLACIES

#5 Simple Forms Develop Into Complex Forms Of Life In Time

According to evolutionary thinking, life began on Earth two or three billion years ago and proceeded through the thousands of generations to develop from so-called "simple" forms into more "complex" forms. The earliest forms of life are often called "primitive," or "lower forms."

Beware! When you read or hear evolution-biased accounts, the words "primitive" and "simple" are deliberately used to convey the idea that evolution has really happened.

THINK! When you realize the incredible design and complexity of even the so-called "simple cell" it becomes very clear that NO LIFE IS SIMPLE!

Questions:

Is life simpler because it is smaller?

Are larger forms more complex than smaller ones?

Is there such a thing as "primitive" fish, frog, or mammal? Or are they all really just as advanced as any other?

What defines primitive? Does the definition imply a gradual progression of evolution?

HOW DID THIS SUPPOSED DEVELOPMENT HAPPEN?

Mutation and Natural Selection are the assumed miracle-workers of this long evolutionary chain of events.

THINK! Aren't mutants deformed or distorted in some way? If someone called you a mutant, you wouldn't exactly take it as a compliment. That's because we all know what mutants are. Why do many seem to forget this when theorizing on evolution?

WHAT IS A MUTATION?

1. Rare in the first place

2. 99.99% are harmful or deadly

WHAT HAVE MUTATION EXPERIMENTS PROVEN?

Fruit flies have been subjected to all manner of mutant-producing experiments in the laboratory. For over 1,500 generations these little bugs have been doused with chemicals, bombarded with radiation and put through the torture test to synthetically produce mutant offspring.

The result?

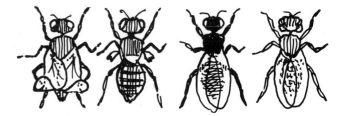

Distorted, damaged fruit flies were produced plentifully.

- *shriveled wings - crooked bodies*

- *weak eyes, no eyes - sterile flies*

But never once has an improved fruit fly been spawned; much less a new kind of creature!

It's no wonder one Nobel scientist has said:

"Trying to improve an organism by mutation is like trying to improve a Swiss watch by dropping it and bending one of its wheels. Improving life by random mutation has a probability of zero."

WHAT ABOUT NATURAL SELECTION?

Is natural selection really the major cause of evolution as popular textbooks so boldly declare?

THINK! Natural selection, in reality, is God's way of picking the best-suited and strongest individuals of a given kind in order to perpetuate the furtherance of that kind.

Inferior oddities are automatically weeded out by natural selection. The best is preserved to maintain genetic stability. What does this tell you would happen to mutants?

THE CLASSIC EXAMPLE: THE PEPPERED MOTH

Many school texts and nature books with a pro-evolution bias have referred to the peppered moth of England as a good example of "evolution in action."

We're told that the light variety of Biston betularia (the peppered moth) evolved into the dark variety during the industrial revolution as a result of factory soot coating the tree trunks where the moths landed.

What really happened?

Actually, it is a known fact that both light and dark varieties of the peppered moth lived in England prior to the industrial revolution. The dark ones were rare. As the light colored

trees darkened with soot, the prevalent white moths landing on the trees became more visible to bird predators. But the rarer dark moths now were better camouflaged and their population began to flourish better than the white ones. Is that evolution? NO, of course not. But why is it deceptively used to suggest a proof of evolution?

WHAT ABOUT DOGS?

Are dogs an example of evolution? There are at least 200 different kinds of dog today. They all trace back their ancestry to an original dog stock several hundred years ago. Actually, this is an example of breeding selectively. It's the same genetic process responsible for the hybridization of cattle or roses. From the lowly Chihuahua to the great St. Bernard, they are all still dogs. Left to themselves for several generations, you'll end up with mongrels. But never once has some offspring turned up with a curious "meow," or a pig snout, or some other totally new and beneficial feature not found in the dog genetic pool.

"GOD GIVES IT A BODY AS IT HAS PLEASED HIM, AND TO EVERY SEED HIS OWN BODY. ALL FLESH IS NOT THE SAME FLESH: BUT THERE IS ONE KIND OF FLESH OF MEN, ANOTHER FLESH OF BEASTS, ANOTHER OF FISHES, AND ANOTHER OF BIRDS."

1 Corinthians 15:38 & 39

Is Life On Earth Actually Getting Better? More Complex?

When you look at the evidence of the history of life on the Earth what do you find? Is life on Earth really getting better or more complex as evolution ideas insist?

Actually **DEGENERATION and EXTINCTION are the rule!** The evidence is well documented and you'll be able to add many good examples to the list presented here.

THE CHAMBERED NAUTILUS

Source: *National Geographic,* January 1976

This fascinating shellfish is explored in *Geographic,* telling of their "progress over the years."

Quoting from the article: "It remains essentially the same as its ancestors of 180 million years ago . . . a living link with the past."

Notice this: "Some 3,500 nautiloid species once flourished. A nine-foot fossil turned up recently in Arkansas." Now, " . . . fewer than half a dozen species still exist . . . time has whittled these descendants to about eight inches."

HOLD IT! Did they say ***"progress?"*** But it's ***"essentially the same."*** Is progress defined by ***"whittling"*** the species from 3,500 to only 6, and cutting down their size from nine feet to just eight inches? Does this evidence convince us of evolution or fit more closely with what we should expect if creation of distinct kinds is true?

The picture of a fossilized bat skeleton has been published in a number of places. It's regarded by experts as the oldest bat remains know, and it supposedly is 50 million years old! (dated by the hypothetical chart of geologic history). The amazing mystery is the fact that the details of the skeleton are virtually the same as those of modern bats!

WHAT ABOUT INSECTS?

Source: *National Geographic,* January 1981

A fossilized cockroach is pictured along with this article about the amazing bug "designed for survival." According to Geographic, the "300-million-year-old fossil imprint . . . shows that roaches have changed but little since their world debut more than 320 million years ago."

> *THINK!* If they are the same today as they've been for that long, stop to think what "their world debut" must have been like when they first came on stage!

If roaches are the same as they ever were, so far as the evidence indicates, then why didn't the bug evolve into some "more complex" form? Yet this is *exactly the evidence to be expected if creation was divinely planned* and creatures really do reproduce "after their kind."

WHAT WAS LIFE LIKE IN THE ANCIENT TROPICAL WORLD?

Source: *National Geographic,* September 1977

In a variety of places on Earth ancient insects have been found encased in petrified tree sap. This amber has a special story of its own that should give us understanding on how things have evolved over the aeons.

A beautifully preserved ant is declared to be *100 million years old.* It's still an ant! And the remarkable thing is that it resembles types of ants living today.

The praying mantis in amber is fully like its descendants living today, yet it's dated by the evolutionary scale at *40 million years old.*

And imagine, if you can, how an ordinary housefly and termite managed not to change into something more desirable in all the "millions of years" since they were stuck in that sap.

KEEP EXPLORING

Some writers have called the phenomenon we're examining, "fixity of kinds." Though there are minor "changes" that do occur within kinds through time, there are definite boundaries. That's why we do not find one kind of creature shading gradually into another kind.

Keep a record of other reports you read about, verifying this principle.

My list of other evidences that kinds have stayed essentially the same and have not evolved into other kinds:

KIND:	SOURCE:	NOTES:

Unlocking the Mysteries of **CREATION**

Do Fossils Show A Gradual Transition Of Evolving Animal Kinds?

HAVE FOSSIL LINKS BEEN FOUND?

If evolution were true, links between kinds should be found plentifully. Paleontologists have been digging fossilized remains of animals out of the Earth for years now, but what have they found to prove evolution? Darwin himself predicted that abundant fossilized links would be found showing transitions from one kind of creature to another.

THINK! If evolution were true why don't we see all manner of living transitional kinds?

Have we ever found a fish emerging from the water to become the first amphibian? You've seen the pictures but has such a creature ever really been seen?

NO! And in case someone refers you to the oriental mudskipper as a supposed proof, help them keep in mind that it is still a fish!

WHALES WITH LEGS?

The evolution of whales has been quite a mystery that has evoked some rather imaginative scenarios. Supposedly, a land-dwelling mammal somehow gradually made it back to the sea, gave up his legs for a spout and a few other things, and in time, the sea monsters of all time just "evolved."

One recent article from a popular science magazine reported the conclusion that one paleontologist has drawn from his recent "find." [2]

According to Philip Gingerich, whales may have walked on four legs 50 million years ago. He found a skull and several teeth and came to the remarkable conclusion that they had belonged to an ancient walking whale. The article draws some interesting associations as you'll see.

"How did he know that the piece of skull and teeth had anything to do with a whale, let alone a walking whale?

"Clue: the sediment in which the skull was found had been either a seashore or a riverbank. Conclusion: the creature lived in or near the water. Clue: the teeth were almost identical with those from known primitive whale fossils found on the west coast of India. Conclusion: the creature was probably a whale."

Watch carefully as the "scientific" conclusion to the article is drawn.

"It must have been heading in the direction of modern whales but wasn't quite there. So its legs probably had not yet evolved into flippers. Most of all though he (Gingerich) *hopes to find leg bones* belonging to the whale species. 'It is possible that we will find some,' he says, 'but **we will be lucky if we do.'**"

You draw your own conclusions about the logic of that.

THE SUPPOSED EVOLUTION OF BIRDS

The remains of an ancient bird, caught in flooded silt and preserved, have been called *Archaeopteryx.* For years it has been proposed as a link between reptiles and birds.

Why would some believe this to be a link when it is fully feathered and with a bone structure which can be defined as true bird?

As Dr. Duane Gish points out in *Evolution: The Fossils Say No!,* the only reason some make a correlation between reptiles and Archaeopteryx is because of its clawed wingtips and a beak full of teeth. Dr. Gish brilliantly points out the truth about this fossil. Some living birds today have claws on their wingtips, yet they are 100% bird. Teeth in an ancient bird say nothing of its ancestral connection to lizards. Since 1984 bird fossil discoveries in Texas have been given ages by evolutionists "millions of years" older than the dates assigned to Archaeopteryx! And these are fully modern birds.

SO HOW DID BIRDS EVOLVE?

Some evolutionists believe that birds evolved their ability to fly from the trees down.

Look up the *1980 Science Yearbook.* According to a feature article there, "paleontologists assumed that the bird's ancestors (which by the way have never been found) learned to climb trees to escape from predators and to seek insect food. Once the 'bird' was in a tree, feathers and wings evolved (as if by magic) to aid in gliding from branch to branch."

Others have different ideas.

According to some "experts" the primitive grandmother of all birds "ran along the ground chasing flying insects which they nabbed with their teeth or front legs! Longer feathers on the front legs then evolved to act as an insect 'net,' and so the legs simply became wings. Then they used the wings to make flapping leaps after insects."

As the report concludes: "If Archaeopteryx did not fly, it was just on the verge." Controversial investigations by British evolutionist, Sir Fred Hoyle, now cast suspicion that Archaeopteryx was fraudulently fabricated by its "discoverers" in the mid-1800's. Even though we have zero evidence for the evolution of birds, some popularizers and museums today are saying that modern birds are really feathered dinosaurs ... highly evolved of course!

National Geographic, August 1978, reported the conjecture of one scientist who proposes that a creature like this evolved into modern birds!

Some Famous Fossil "Connections" To Evolution

There are many places on Earth where fish fossils have been found. For years the fish called *"Coelacanth"* (pronounced SEE-la-kanth) was labeled as a "transitional form." It was supposed to be an evolutionary "link" between fish and amphibian.

The Coelacanth was presumed to have been extinct for the last 70 million years. Consequently, it was used as an "index fossil" to ascribe ages to geologic layers. If you happened to find one of these petrified skeletons you "knew" you were into some 70 million year old rock.

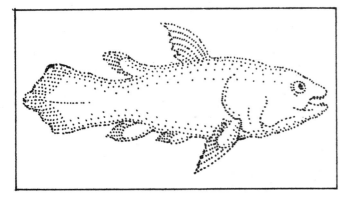

A variety of certain extinct animals have been age-designated by evolution-believing paleontologists. They believe that so-called "lower forms," which are now extinct, fit into a certain time sequence of millions of years. They automatically assume that fossilized remains of them will provide an "indexing" of the age of the rock layers in which they are found.

A SURPRISING MYSTERY

In the 1930's **fishermen hauled in a live Coelacanth off** the coast of Madagascar. It was picked up by a woman in the fish market and turned in to a scientist. He properly identified it as a genuine Coelacanth.

Since then, as many as thirty specimens of Coelacanth have been caught alive. Reports are common in the literature dealing with fossils. It is often billed as a **"living link"** with the past. But rarely are people made aware of the important implications of this so-called "primitive" living fish.

The Coelacanth is obviously not an evolutionary link at all! It did not become extinct 70 million years ago. And if you were to discover a fossilized Coelacanth it would tell you nothing about the age of the rock in which you found it. Again, the evidence supports the concept that kinds remain essentially the same. There is no evolution of kinds from other kinds.

THE EVOLUTION OF HORSES: FACT OR FICTION?

The so-called "horse series" of transitional forms has been widely taught and published. Biology teachers have been given this convincing looking chart to demonstrate to their classes a clear example of evolution.

What are the facts?

The creatures represented in the chart are based on bones that have been found in widely diverse locations: India, North America, South America, and Europe.

The creatures have been cleverly arranged in the chart with the small ones at the bottom and the large ones at the top. This does indeed make the order look like a progressive evolution, doesn't it?

Some of the specimens were questionable. Some felt they should be included as links to hippopotamus or some other animal, rather than to the modern horse.

In some places on Earth two or more of the horse series have been found buried together in the same layers of rock. This indicates they lived at the same time and were not distantly related transitions. In Florida the fossil bones of seven different kinds of horse have been found together in the same cataclysmic burial place. In one South American location the order of the chart is apparently upside down. The modern horse is deeper, and the more "primitive" three-toed type is higher up in the rock strata.

Another mystery seldom told is that the specimens in the chart have an unusually illogical evolution of the rib cage. Beginning at the bottom, the earliest and smallest creature has 18 pairs of ribs. The next one up has 15. Then comes a "higher" horse with 19, but then it drops down to 18 pairs at the top of the chart. There doesn't seem to be any continuity of evolution displayed in this.

These kinds of findings have led some scientists to recognize the difficulties of accepting such "evidence" for evolution.

Dr. David Raup, curator of the Field Museum of Natural History in Chicago, writes this in an article of the museum's *Bulletin* [vol. 50(1), 1979, pp. 22-29].

"Classic cases ... such as evolution of the horse in North America have had to be modified or discarded as the result of more detailed information."

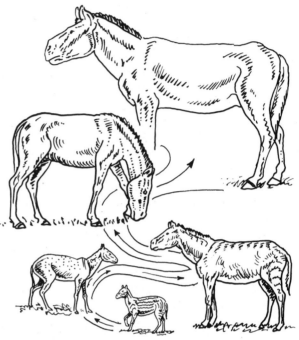

PUZZLING TO DARWIN

In his book, *Origin of Species* (1859), Darwin admitted: "As by this theory innumerable transitional forms must have existed, why do we not find them embedded in countless numbers in the crust of the Earth? The number of intermediate links between all living and extinct species must have been inconceivably great!"

As *Newsweek* magazine pointed out recently (11/3/80): "Missing links are the rule. The more scientists have searched for the transitional forms between species, the more they have become frustrated."

Stephen Gould of Harvard boldly tells it like it is in *Natural History,* vol. 86(5), 1977, page 13.

"The family trees which adorn our text books are based on inference, however reasonable, not the evidence of fossils."

Does The Geologic Chart Prove Evolution?

WHAT IS THE GEOLOGIC CHART?

Devised in the early 1800's, long before carbon 14 was discovered, this common chart of Earth's supposed history is totally geared to the "theory" of evolution. It diagrams over a half a billion years of impressive rock layers in which are seen the fossilized remains of creatures showing a supposed "picture" of how evolution is verified by the rocks of the Earth.

NO SUCH PROGRESSION OF ROCK LAYERS AND CORRESPONDING FOSSILS IS FOUND ANYWHERE ON PLANET EARTH!

HOW ARE FOSSILS DATED?

When fossils are found and displayed, the dates you hear about are based on this chart. It doesn't matter if it's a fossil clam, fish, or dinosaur, the age for the specimen is determined by the chart.

LET'S SEE HOW IT "WORKS"

If you found some fossil trilobites and want to know their age you naturally take them to the paleontologist at your local university. He will ask you where you found them. Then he will identify the layer as "Cambrian." Next he will take you over to The Geologic Chart on the wall. "Here it is on the chart," he'll say. "So how old is that layer?" you ask.

He replies, "About 450 million years."

In your naive curiosity you ask, "How do you know that?"

His professional-sounding response, sounding slightly intimidating, is, "Because they are in the Cambrian layer, of course."

Now you're challenged to get to the heart of all this, so you walk across the hall to see the geology professor. You show him a picture of the place you found your fossil trilobites. You don't want to confuse the issue with all you've just learned, so you simply ask him, "Sir, how old would you judge this layer of rock to be?"

"That's easy," he says, "What kind of fossils have you found in it?"

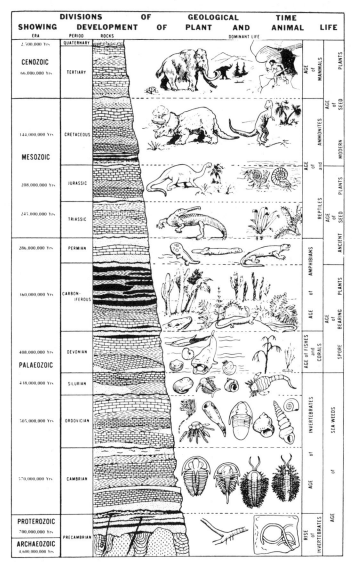

"Trilobites," you confidently reply.

"Easier still," he says. "It's about 450 million years old."

Wanting to be doubly sure of the facts, you inquire, "How did you know that?"

Possibly a bit indignant at your revealing investigation, he tells you, "If trilobites are there then we know the layer is that old."

Cambridge University's Geologist, R.R. Rastall, wrote in *Encyclopaedia Brittanica* (1956, vol. 10, p. 168): " . . . geologists are arguing in a circle. The succession of organisms has been determined by the study of their remains imbedded in the rocks, and the relative ages of the rocks are determined by the remains of organisms they contain."

LIVING FOSSILS?

A little sea creature called Apus has been labeled a "living fossil" because of its remarkable similarity to the trilobite. Why didn't these creatures phase out altogether as evolution marched on to greater things? Some feel that since trilobites lived in the depths of the ocean floors it is entirely possible to find living trilobites. There goes another "index fossil" if they are found!"

WHAT ELSE HAS BEEN LEARNED AT THE BOTTOM OF THE CHART?

The October 9, 1980 issue of the *Vancouver Sun* reported about the Burgess shale beds at Field, B.C., known famously for the trilobites found there. But there is a vast deposit of complex animals found there as well.

Quoting from the article, "Animals . . . are so bizarre in this massive graveyard that not even the most imaginative paleontologist has been able to connect them to any family of modern animals."

And note the "expert" conclusion: "The **textbook evolutionary tree,** with everything traced back to a few common ancestors, **is inaccurate!"**

Does this mean that typical museum displays and textbook charts showing the familiar molecule to man idea are misleading?

A BIGGER MYSTERY FOR EVOLUTION

From Utah has come a rare and peculiar find. A man named William Meister split open a slab of shale only to discover a clear impression of a **human** sandal print. But the real surprise was to see a good fossilized **trilobite** embedded in the heel of the sandal print! [3]

MAJOR MYSTERY OF EVOLUTION?

Besides trilobites it has been repeatedly confirmed that the "Cambrian" layer is loaded with every major invertebrate form of life including sponges, corals, worms, crustaceans, jellyfish, and mullusks. Billions of fossils of well-developed marine organisms are found in this "basement layer of life" worldwide. Below that layer, in what has been called Pre- Cambrian, there is **a striking absence of life forms.**

simplified cross section of geological layers shows that fossils appeared suddenly in the Cambrian period.

Geologist H.S. Ladd, in the Geological Society of America Memoir 67 (1957) wrote: "Most paleontologists . . . (are) ignoring the **most important missing link of all.** Indeed the missing Pre- Cambrian record cannot properly be described as a link for it is in reality, about nine-tenths of the chain of life: **the first nine- tenths!"**

It is because of this problem that the famous evolutionist George Gaylord Simpson has said that the absence of Pre-Cambrian fossils is **"the major mystery of the history of life."** [4]

The fossils of the Earth reveal exactly what you would expect if the creation explanation is true. Multiplied living kinds appear abruptly with no "primitive" links leading up to them.

With all these "mysteries" to cloud the issue of evolution, some expert evolution proponents have escaped to new realms of "reason."

PANSPERMIA

Dr. Francis Crick, the famed biologist and Nobel prize winner for his work with the discovery of DNA, has proposed a "superman slant." He says, " . . . that life on Earth may have sprung from tiny organisms from a distant planet, sent here by space ship as part of a deliberate act of seeding." [5]

But he fails to say anything about life being "planted" here by God!

HOPEFUL MONSTERS FOR WISHFUL THINKERS

There's always the "hopeful monster" theory to dodge behind. A scientist named Goldschmidt proposed it years ago. Supposedly two reptiles mated; the female laid her eggs; and when one of them hatched out popped something like a parakeet! The same approach is suggested for every one of the kinds of life on Earth since **no genuine transitional forms have ever been found.** Recently this idea has been revived under a new name: "punctuated equilibrium." Now, that does sound more "scientific" doesn't it? The illusionary idea is that creatures stay the same until "something irregular" PUNCTUATES the status quo with sudden "leaps" of evolutionary change (which nobody ever witnessed)."

MANY SCIENTISTS ARE IN A QUANDRY ABOUT DARWIN'S THEORY

An issue of *Newsweek* magazine recently (4/8/85) carried an article titled "Science Contra Darwin." Unknown to much of the general public, the controversy over the origin of life is not just an argument between conservative Christians and hard-nosed materialists.

The L.A. Times reported (6/25/78): "Scientists behave the way the rest of us do when our beliefs are in conflict with the evidence. We become irritated, we pretend the conflict does not exist, or we paper it over with meaningless phrases."

WHAT IS THE REAL ISSUE?

The apostle Paul said it plainly in Romans 1:21. *"They became vain (empty) in their imaginations (speculations); their foolish heart was darkened; professing themselves to be wise (intellectual arrogance) they became fools (depraved and deluded)."*

"THEY EXCHANGED THE TRUTH OF GOD FOR A LIE!" (Romans 1:25)

When things don't add up, you either have to cover it up or change your mind. It's no wonder we're hearing some rather bizarre explanations nowadays. We should expect it.

The wisest man who ever lived (King Solomon) wrote:

"A fool's speech brings him to ruin. Since he begins with a foolish premise, his conclusion is sheer madness."

Ecclesiastes 10:12-13

> *THINK!* Are the ridiculous escapes from reason just conclusions based on a foolish premise?

LET US ALL REALIZE THE REAL PROBLEM

Scientist D.M.S. Watson wrote years ago:

"The theory of **evolution** is universally **accepted** not because it can be proved by logical, coherent evidence to be true, but **because the only alternative, special creation, is clearly incredible!"** (*Nature*, vol. 124, p. 233, 1929)

"Clearly incredible?" We all need to be bold enough to ask: "Which of these considerations is really incredible?"

LIFE FROM CLAY?

Now some scientists are saying that life evolved from clay! Isn't it amazing that some can not separate the vast difference between the natural crystalizing properties of minerals and the infinite complexity of the living cell?

People used to think the Earth was flat and the center of the universe. Until less than a century ago respected doctors practiced "bloodletting" to treat sick patients. What has happened? Scientists HAVE been proven wrong on occasion.

Amazingly, leading evolutionist astronomer Robert Jastrow indicated that it will ultimately be necessary for science to come to terms with the supernatural. As he puts it: **"For the scientist who has lived by his faith in the power of reason, the story ends like a bad dream. He has scaled the mountains of ignorance; he is about to conquer the highest peak; as he pulls himself over the final rock, he is greeted by a band of theologians who have been sitting there for centuries."** (*God And The Astronomers, by Jastro, 1978*)

"SEE TO IT THAT NO ONE TAKES YOU CAPTIVE THROUGH PHILOSOPHY AND EMPTY DECEPTION, ACCORDING TO THE TRADITION OF MEN, ACCORDING TO THE ELEMENTARY PRINCIPLES OF THE WORLD, RATHER THAN ACCORDING TO CHRIST."

Colossians 2:8

Evolution Opposes Christianity: There Is No Compromise

- *Evolution contradicts the Bible record of a finished creation.*
  - The creative process is not now operating. God rested. (Genesis 2:1-3)
  - Evolution supposes a continuous creative process.

- *Evolution contradicts the doctrine of fixed and distinct kinds.*
  - Evolution supposes a constant change back to a common ancestor.
  - All flesh is not the same. 1 Corinthians 15:38 & 39
  - Biogenesis (life begets life) is seen as "like begets like."

- *Evolution is inconsistent with God's omniscience.*
  - It is all chance and errors in genetic translation to successive generations.
  - God is orderly and not the author of confusion. 1 Corinthians 14:33

- *Evolution contradicts the universal principle of decay.*
  - Cursed is the ground (Genesis 3:17)
  - The creation is waiting for release from death. (Romans 8:21)
  - The law of degeneration, disintegration, and increasing entropy is universally observed.
  - The molecule to man philosophy is an illusion that ignores all the genuine evidence of nature.

- *Evolution Produces anti-Christian results.*
  - A corrupt tree cannot produce good fruit. (Matthew 7:18)
  - Evolution is at the very root of atheism, communism, relativism, racism, anarchism, and all manner of anti-Christian practices.
  - Dangerous and deadly social problems are deeply rooted in the purposelessness of materialistic evolution: suicide, promiscuity, abortion, and chemical abuse, just to name the most obvious.
  - Evolution has played a major part in reducing how humans think of themselves. Obvious problems are: low self-esteem, animalistic behavior, and depression due to a feeling of meaninglessness in life.

"THERE IS A WAY WHICH SEEMS RIGHT TO A MAN, BUT ITS END IS THE WAY OF DEATH."

Proverbs 14:12

SUPPLEMENT
TO SESSION TWO

EXPLORING THE DESIGN AND PERFECTION
OF THE CREATION

*"But now ask the beasts, and they will teach you;
and the birds of the air, and they will tell you;
. . . Who among all these does not know
that the hand of the Lord has done this,
in whose hand is the life of every living thing,
and the breath of all mankind?"*
Job 12:7, 9, 10

NATURE'S CHEMISTS?

In the 1981 edition of *World Book Encyclopedia's Science Yearbook* a fascinating article begins:

"The bombardier beetle is one of many insects that makes and uses complex chemicals to protect itself from its natural enemies."

Upon examining this creature in the laboratory, biologists have discovered that inside the bombardier beetle's body are two special chambers which produce two special chemicals. When these chemicals are mixed and ejected through a special tube strategically positioned at the rear of the beetle's body, they produce a virtual explosion in the face of would-be attackers.

The hapless predator is left gagging in a steaming, noxious spray that issues forth at the boiling point of water.

His special turret-like artillery fires with accuracy in whatever direction is necessary.

A very special feature of this lowly "chemist" is another chemical called an "inhibitor." If you were to mix the basic two ingredients of his chemical warfare (hydrogen peroxide and hydroquinone) it would blow up in your face. But the addition of the "inhibitor" prevents such a premature blast. Only when the "bomb" is ejected does it blow up for the bombardier beetle's defense.

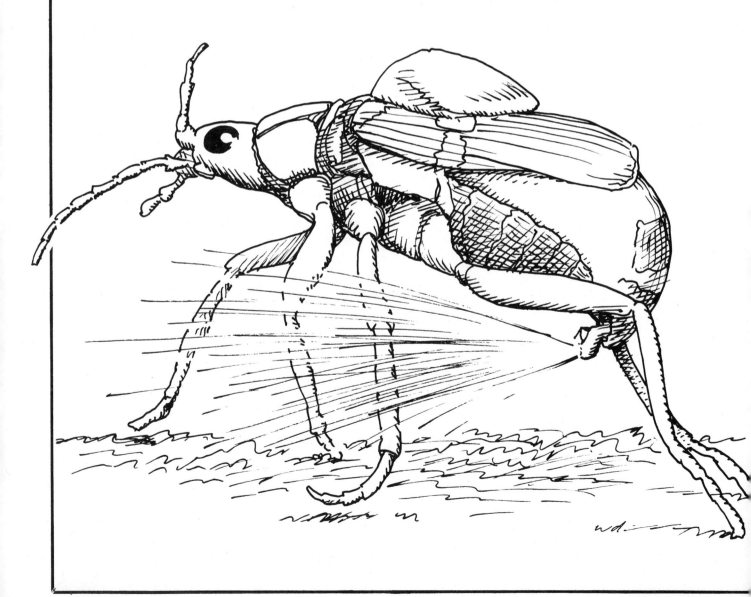

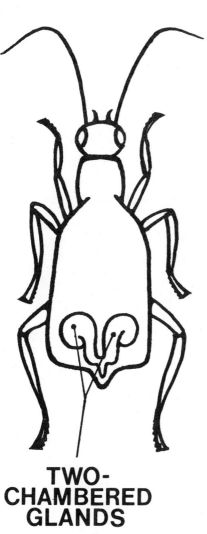

TWO-CHAMBERED GLANDS

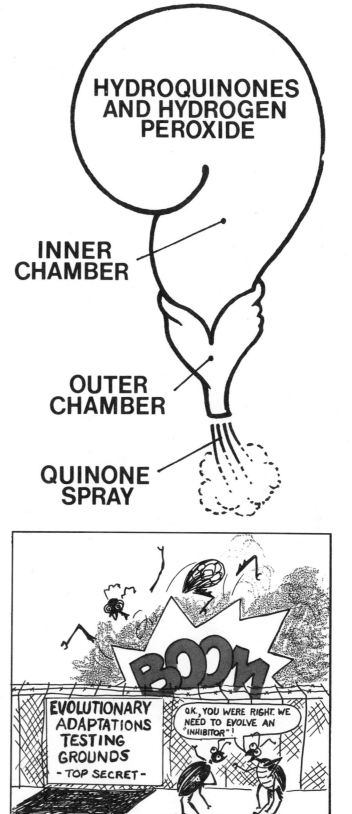

HYDROQUINONES AND HYDROGEN PEROXIDE

INNER CHAMBER

OUTER CHAMBER

QUINONE SPRAY

WITNESSING A REAL MIRACLE

Try to imagine the extremely delicate defense system of the bombardier beetle before it had evolved to be fully functional.

If the chemicals were not just the right strength or of just the perfect ingredients, can you see the picture of Mr. Bombardier Beetle manning his battle station for the first time? If it fizzled he would be exterminated in short order by his predators for inadequate protection.

What about another possible hazard?

If the inner chambers or tubes weren't perfectly organized from the beginning, or if the inhibitor technology was not quite ready in time, think what would happen. He goes to take a deadly shot and it backfires! Mr. Bombardier Beetle is blown to pieces and never even makes it to the endangered species list.

EVOLUTIONARY ADAPTATIONS TESTING GROUNDS
- TOP SECRET -

O.K., YOU WERE RIGHT. WE NEED TO EVOLVE AN "INHIBITOR"!

You don't have to have a college degree to realize that a multitude of precision details had to be working perfectly from the beginning of this bug's existence.

Unlocking the Mysteries of **CREATION**

The Wonderful Woodpecker

The woodpecker is totally different from other birds. Every part of his body is especially fitted for drilling into wood.

His short legs and powerful claws are absolutely essential for holding on tightly to vertical tree trunks. They are very different from the spindly legs of other birds.

The beak of the woodpecker is also very unique. Banging away as much as a hundred times a minute, the woodpecker's beak has to be much harder than the beaks of other birds. Even more special is the way his beak is connected to his skull. There is a resilient shock- absorbing tissue in between the beak and the skull that is not found in any other birds!

Or consider the woodpecker's tongue. This most remarkable instrument is barbed in most of the 179 species of woodpecker found in the world. It's about four times longer than the beak as it wraps around the back of the bird's skull in a very purposeful design. Some woodpeckers produce a sticky substance coating the tongue for baiting ants. All of the

woodpecker family use this snake-like tool for penetrating deep inside a tree trunk to ferret out ants and grubs.

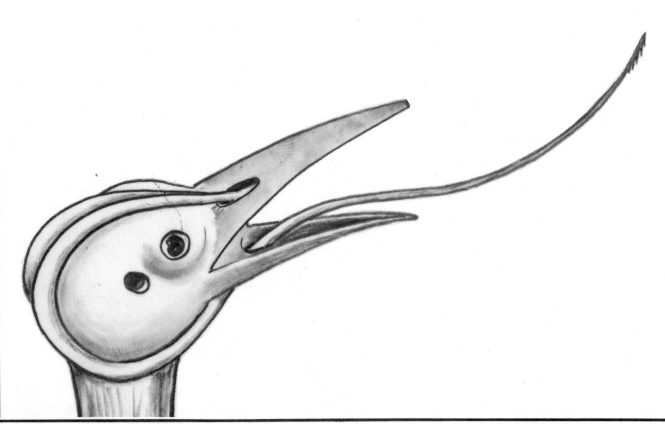

AN EXACTING ENGINEERING ACCIDENT?

After realizing just a few of the marvelous details of our little feathered "machines," try to imagine how such an invention could be engineered by a collection of rare and serendipitus chances.

Besides his powerful beak and special tongue, keep in mind that he also has a keen sense of smell. Together with his highly sensitive ears, Mr. Woodpecker can detect insects crawling around under the bark of the tree he's scaling. By the way, how did the mysterious woodpecker learn to climb trees so well?

His specially constructed tail feathers are stiff enough to actually brace him securely wherever he climbs the forest's "walls."

The engineering for such a technological wonder as the woodpecker boggles our minds.

Now, think about it. If the woodpeckers did evolve, can you imagine the obstacles the first one would have had to overcome?

Think of the primeval headaches due to a lack of properly developed cushioning in the beak design! Worse yet, think of all the broken beaks that weren't hard enough to take all the jack-hammering yet!

What about the primal woodpecker's legs and claws? In the early days, how on Earth did the poor thing manage to hang on for dear life, risking everything for the high- class meals at the top? You can't help but wonder why would this primitive dare-devil bother to peck away for grubs in wood of all places? After all, there have always been plenty of bugs crawling around on the dirt below.

It becomes quite apparent, even to the casual bird-watcher, that no supposed ancestral mutant would have ever survived to produce the marvel of God's creative genius we so commonly see today.

Unlocking the Mysteries of CREATION

The Graceful Giraffe

When you take just a look at the crane-like giraffe you can't help watching that amazing neck as it moves up and down.

The giraffe has the highest blood pressure of all animals. When you consider what a stretch uphill it is to pump the blood to his head you can appreciate the problem.

But even more interesting are the intermittent valves in the blood vessels of the long neck. An amazing demonstration of plumbing technology, these valves are vital to our spotted tropical friends. When the giraffe lowers his head for a drink the valves quickly close down to keep the massive blood supply from rushing harmfully to the brain. When he raises his head the precision-working valves prevent the blood from rushing away from the head too quickly. [6]

Now you know how it is when you have been bending over and then suddenly lift your head up. Think of the giraffe. Imagine those valves not being perfectly functioning at any supposed evolutionary stage. The poor giraffe would die of a cerebral hemorrhage the first time he bent down for a drink. Or the simple act of raising his head up from a nap would result in such a loss of blood from the brain that he would likely pass out and be easy prey for a nearby lioness.

Again, the conclusion is obvious to one willing to discover the truth. As you continue to investigate creations as humble as the little honeybee or as grand as the marvels of our human bodies, the evidence of intricate design is everywhere.

"O LORD HOW MANIFOLD ARE THY WORKS. IN WISDOM HAST THOU MADE THEM ALL. THE EARTH IS FULL OF THY RICHES."

Psalm 104:24 (KJV)

"The things that are made" tell us about the Creator who is the all-powerful designer of them all. (Romans 1:20).

That is why the Bible declares without apology, *"they are without excuse."* Think what that means. No one will ever be able to stand before God and tell Him there is a good excuse for not believing in and obeying the call of God. Every human who ever lived has had the clear evidence of the designs and intricacies of creation to loudly declare to him, "There is an Almighty God who made all this."

And as Jeremiah the prophet said,

YOU SHALL SEEK ME & YOU SHALL FIND ME WHEN YOU SEARCH FOR ME WITH ALL YOUR HEART

© 1987 PRINTS OF PEACE

(29:13)

Unlocking the Mysteries of **CREATION**

THE WORKS OF THE LORD ARE GREAT,
STUDIED BY ALL WHO HAVE PLEASURE IN THEM.

Psalm 111:3

References For Section Two

There is a great deal of data in this section, most of which can be found in a number of excellent source books dealing with the creation vs. evolution controversy. Rather than try to document every detail, I refer you to the selected reading list below. Since many of the reference sources are incorporated in the text, they will not be duplicated here. d.p.

1. Keith, Sir Arthur, quoted in Science and Scripture, v. 3 1971, p. 17.
2. Science Digest Magazine, Nov.-Dec. 1980, p. 25.
3. Cook, Melvin A., "William J. Meister Discovery of Human Footprints with Trilobites in a Cambrian Formation of Western Utah," in Why Not Creation?, Presbyterian and Reformed Publishing Co., Philadelphia, 1970, p. 185-193.
4. Simpson, George G., The Meaning Of Evolution, Yale Univ. Press, New Haven, 1949, p. 18.
5. Bible-Science Newsletter, May 1974, p. 6.
6. Kofahl, Robert E., and Segraves, Kelly L., The Creation Explanation, Harold Shaw Publishers, Wheaton, 1975, p. 5.

Selected Reading

Gish, Duane T., Evolution: The Fossils Say No, Creation-Life Publishers, San Diego, 1978.

Morris, Henry M., The Remarkable Birth Of Planet Earth, Bethany Fellowship, Minneapolis, 1972.

Morris, Henry, Men Of Science, Men Of God, CLP, 1982.

Thaxton, Charles, The Mystery Of Life's Origin, Philosophical Library, New York, 1984.

McDowell, Josh, Reasons Skeptics Should Consider Christianity, Here's Life Publ., San Bernardino, Ca, 1981.

Denton, Michael, Evolution: A Theory In Crisis, Adler & Adler, Bethesda, Md, 1986.

Taylor, Ian T., In The Minds Of Men, TFE Publ, Toronto, 1984.

Sunderland, Luther, Darwins Enigma, CLP, 1984.

SESSION 3

UNLOCKING THE MYSTERIES OF ORIGINAL MAN AND THE MISSING LINKS

"I WILL GIVE THANKS TO THEE

FOR I AM FEARFULLY AND WONDERFULLY MADE;

WONDERFUL ARE THY WORKS,

AND MY SOUL KNOWS IT VERY WELL."

Psalm 139:14 (NAS)

Contents Of Section Three

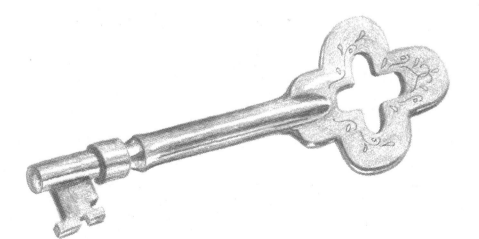

"So God created man in His own image;
in the image of God He created him;
male and female He created them."
Genesis 1:27

IN
THE
IMAGE
OF
GOD

THE BIBLE MODEL OF ORIGINAL MAN

"So God created man in His own image; in the image of God He created him; male and female He created them."

Genesis 1:27

"And the LORD God formed man of the dust of the ground, and breathed into his nostrils the breath of life; and man became a living being."

Genesis 2:7

According to the Bible the fact of man's creation is clearly the result of a sovereign act of the almighty Creator. There can be no mistake that the wording of the scripture and the understanding of conservative Bible scholars down through the ages agree that man's origin was supernatural and finished at his beginning.

WHAT IS THE POPULAR CONCEPT OF MAN'S ORIGIN?

According to the prevailing theory of human origins, modern man gradually evolved from brutish "cave men." These ancestors, primitive as they were, had somehow evolved from some species of "pre-man" over an immense period of millions of years.

A typical explanation of how this supposedly happened can be found in many colorful books.

"Over ten million years ago a versatile monkey sired two distinct lines, the forest apes, and cave-camping pre-humans such as Australopithecus (literally meaning 'southern ape'). One of many branches of the Australopithicenes survived to become true men like Peking Man, a probable precursor of modern orientals."

WHAT IS THE TANGIBLE BASIS FOR THE BELIEF IN HUMAN EVOLUTION ?

The search for fossil evidence to prove the theory of human evolution has gone on for over a century. It has been a miserable failure!

T. EVANS

In the first place, any bones of ancient "prehistoric" men are extremely rare. Yet, as we discovered in an earlier section, it would seem natural to find many burials of early man if human kind has inhabited Earth for a million years or more.

Nevertheless, popular books, magazines, and school texts strongly define man's evolution as a fact!

The artists have colorfully pictured what our ancestral family tree supposedly resembled.

In this section we will go through the familiar chart, examining the names and the evidences used to support the theory.

Notice the explanation given in the colorful *Reader's Digest* book titled *The Last Two Million Years.* It is quite typical of how the idea is presented to the public.

> **"In Darwin's time there was little evidence to support his theory; but since then a whole chain of 'missing links' has been established by study of fossil bones found at prehistoric sites. The chart . . . shows how, over 40 million years, descendants of the early primates gradually evolved to produce modern man."**

THINK! How can "a whole chain" be "established" when the links are "missing"?

THINK! How can any chart "show how" anything "gradually evolved" when the premise of evolution is still admitted to be only a theory?

And notice how 40 million years is rattled off as though it was unquestionable scientific fact! It appears that writers of such statements are, to say the least, overstating their case. But read on.

> **" . . . the first breakthrough came when creatures adapted to standing and walking in an upright position . . . Most important, they were now able to make and use tools . . . distinguishing [them as] true man . . . As man's brain became bigger, responding to the demands of more complex hunting, he became taller, with more refined teeth and jaws."**

THINK! Knowing that some animals have the ability to use tools in some rather ingenious ways, why would we believe this is the trait "distinguishing true man?"

THINK! What on earth has the challenge of a complex hunting project got to do with the development of a larger brain, or height, or teeth and jaws?

— *Where Are The Missing Links?* —

THE EVOLUTIONARY ADAM

In the chart found in the book, *The Last Two Million Years,* we find the first entry labeled simply, **"Common Ancestor."**

The caption goes on to say: "This creature is believed to have been a forest-dwelling creature, the ancestor from which modern apes and man both descend. ***No traces of such a creature have yet been found.***"

Other charts include the name "Gigantopithecus" as first in line. However this creature has been discovered to be nothing more than an extinct ground ape. In fact, Richard Leakey, the famed anthropologist, has taken it off the chart leading to man and placed it on a totally separate line. One place to see this is in the *Time* magazine of November 7, 1977.

If Leakey, one of the leading evolutionists, has taken what was once thought of as our "common ancestor" off the chart, we might as well remove him too. Since this link has never been found we can only conclude that **he must still be missing!**

RAMAPITHECUS

Next in line is an individual which has been given the name Ramapithecus. It is described as "a more advanced primate... appearing by [sic] 14 million years ago."

The interesting thing about this entry in the chart is that it was reconstructed on the basis of one skeletal fragment: a piece of jawbone about two inches long! The "find" was made in India in the 1930's. Some time later another small piece of jawbone was dug up in Africa and identified as coming from the same species.

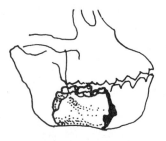

Even though the evidence is fragmentary it appears that some people can make a little of anything go a long way! All the physical features of Ramapithecus have been artistically detailed and published widely: his height, posture, length of limbs, shape of head and amount of body hair.

And just think; all of this from a piece of jawbone two inches long!

"Why include this in the chart?" you may ask. A paleontologist from Yale University by the name of David Pilbeam "believed" this was more man-like than ape-like! But now evidence has shown that some living baboons have similar tooth and jaw structures.

It isn't surprising that this statement was made in *Science Digest* in April of 1981 (page 36).

"A reinterpretation of . . . (this) . . . jaw-. . . now suggests that Ramapithecus was an ancestor of neither modern humans or modern apes. Instead Pilbeam (himself) thinks it represents a third lineage that has no living descendants."

So now we have to remove another "link" from the chart. It's still missing!

AUSTRALOPITHECUS

One of the long-standing and well-known creatures in the familiar chart is known as Australopithecus. The name literally means "southern ape." That should tell you a lot!

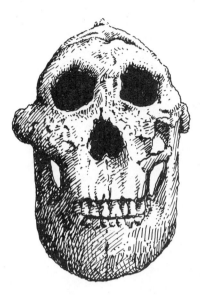

A number of skulls have been found over the years. The most famous in our day is no doubt the find of Dr. Louis Leakey in Africa which he called Zinjanthropus bosei. Two types of this extinct ape are included here: A. africanus and A. robustus.

Since the brain case and skull form of this animal is distinctly ape you may wonder why it is included in the chart of man's ancestry.

The reason for his inclusion is simply this: tiny supposed stone tools were found nearby. When you notice how small this remarkable bit of *technology* is, you would have good reason to question the validity of using evidence like this.

It is interesting that Richard Leakey, in his book called *Origins,* has removed Australopithecus from the chart leading to Homo sapiens. The southern ape has been placed in a totally separate lineage altogether.

If Leakey, one of the leading evolutionists in our day, has removed this supposed "link" from the chart, we may as well remove it too.

Another find deserving mention here was made by a young bone digger named Donald Johanson. His African discovery was featured in *National Geographic* magazine in December 1976. This assortment of bones is claimed by its discoverer to have walked upright.

The interesting thing is that the bone structure of this creature which Johanson named "Lucy" is no different than some modern chimpanzees that walk upright! It's clear that modern chimps are not man's ancestors, so why would anybody suppose that this long dead specimen has anything to do with man?

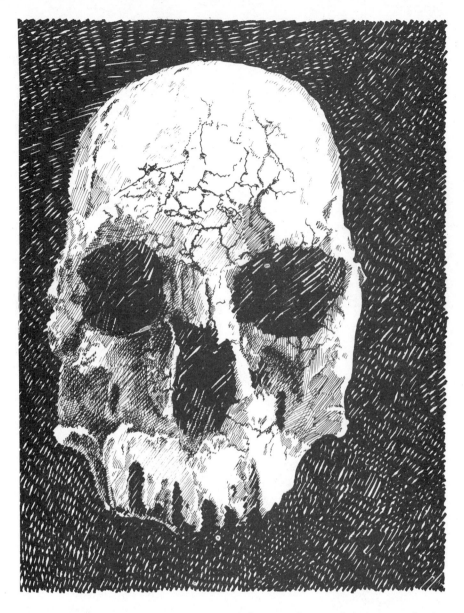

Homo habilis is next on the chart. In June 1973 the *National Geographic* magazine published an article that was devastating to conventional ideas about human evolution. It reported a new find in Kenya, Africa by Richard Leakey, the leading expert on prehistoric man. The find was dubbed "Skull 1470" (Fourteen-seventy) for its catalog number in the Kenya museum.

Leakey made an astounding challenge, highlighted prominently in bold letters by *National Geographic.*

"Either we toss out this skull or we toss out our theories of early man."

The anthropologist said this fossil is 2.8 million years old, yet belongs to man's genus. In other words, it was claimed to be more man-like than any of the other near-man relics on the chart. The problem was that the skull was supposedly dated to be more than two million years older than other creatures on the chart which are clearly more ape-like.

It's no wonder Leakey candidly says "It simply fits no previous models of human beginnings." And because of the skull's "surprisingly large braincase," Leakey shockingly admits, "it leaves in ruins the *notion* that all early fossils can be arranged in an orderly sequence of evolutionary change."

Apparently the editors of *National Geographic* felt the prevailing theories of human evolution could stand some shaking.

Do you realize the impact of Leakey's comments? He implies that the chart we've been seeing publicized is a ruined *"notion."* The "orderly sequence of evolutionary change" doesn't rate any better than a notion?

Following this amazing publicity, Leakey lectured in San Diego, California and other places. He has been reported to have explained his convictions that his discovery eliminates everything we've been taught about human origins. He says he has nothing to offer in place of the popular concepts.

Though Leakey says that Skull 1470 is like true man, isn't it remarkable how the artist can flesh out a face he has never seen? Ears, nose, cheeks, chin, and facial hair are details imaginatively recreated to convey an intended conclusion.

What is the implication? Does it not seem that these people are adding monkey-like features to deliberately influence the public to think they have found an ape- man?

Hold it! If Skull 1470 is more advanced than other supposed ape-like links, then doesn't that nullify those "younger" bones as ancestral to man?

Precisely! That's why Leakey insists that his find ruins the traditional theory of human

Artistic reconstruction of "1470 man"

evolution. But he supposes there must be another explanation. What would they think if you could uncover true human remains dating back to the time of the dinosaurs or ever earlier? Just wait and you will read about even more mysterious finds (mysterious only because they do not fit the traditional evolutionary concepts of human beginnings).

Looking at our popular chart again, notice what has happened so far. All three of the characters on the left have been removed for various reasons. The fourth is considered to be practically modern. Where does that leave us?

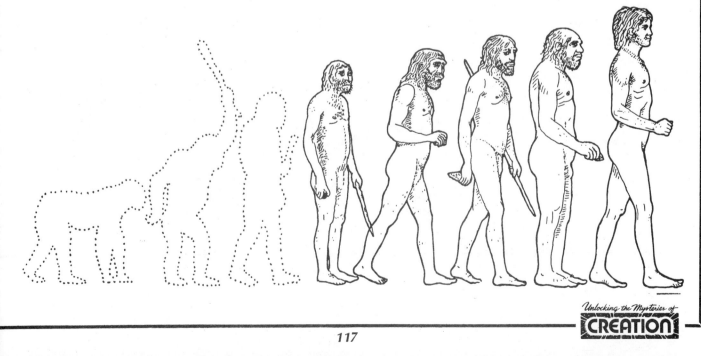

Is Homo Erectus The Missing Link?

In current popular literature, the name Homo erectus is given to bones of what are supposed to be the oldest nearly human remains on Earth. As pointed out in the title for chapter four of the *Life Nature Library* volume on Early Man, Homo erectus is "A true man at last."

But is he? Take a look at the genuine evidence to form your own conclusion and see if the literature you read is telling the full story. The classification "Homo erectus" is based on two fossil finds: "Peking Man" and "Java Man." The designation "erectus" refers to the ability to walk in a fully erect or upright posture, unlike the "pithecuses," which are ape types that use the forelimbs in walking.

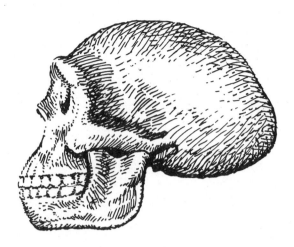

PEKING MAN

In 1921 two molar teeth were dug from a limestone hill 25 miles from Peking, China. Six years later a third tooth was found and given to Dr. David Black. Over the next several years dozens of pieces of broken up skulls were found. Some time after 1936 a man named Franz Weidenreich, who was in charge of the dig, fashioned a model of what "Peking Man" supposedly looked like.

Unfortunately, during World War Two, **all the fragmentary evidence except for two teeth was lost.** So there is no way to subject Peking Man to critical modern analysis.

French scientist, Marcellin Boule, examined the actual fragments of skull, and in 1937 published his opinion that the find was decidedly monkey-like. Boule and others report that **the model did not correspond objectively to the fossils.** It was clear that the fragments of skull found belonged to creatures hunted by true humans.

The age ascribed to Peking Man is supposed to be in the neighborhood of half a million years. But the dating of the site appears quite inconclusive. **Human fossils have been excavated from the same site!** Plant and animal remains are similar from top to bottom of the 160-foot deep dig. Man-made tools are also found there. And there is also evidence that a limestone processing industry was operated by the supposedly stone age people buried at the site. Dr. Duane Gish documents the background of the whole story very well in his book, *Evolution: The Fossils Say No.*

THINK! What kind of science is it that depends on models of evidence that can't be found?

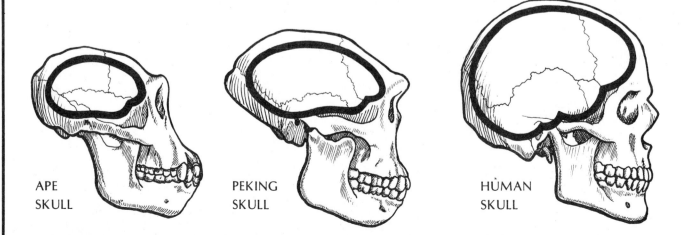

APE SKULL   PEKING SKULL   HUMAN SKULL

JAVA MAN

In recent years claims have been made that Homo erectus fragments have been found in Africa. However, besides Peking Man, the only other significant erectus find comes from the South Pacific island of Java.

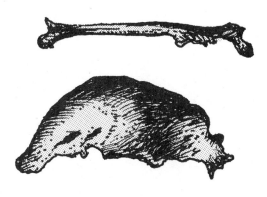

In 1891 a Dutch physician named Eugene Dubois discovered "Java Man." Well, what he really found was an ape-like skullcap.

A year later he returned to the site along the Solo River and found a human leg bone and two molar teeth 50 feet away from the first find.

Like a good scientist, Dubois put the head bone together with the leg bone and called it "Java Man."

Expert evolutionists have ingeniously estimated the age of the find to be 500,000 years old.

Dubois deliberately set out to find the missing link in Java. But when he shared his "evidence" with the experts of his day they weren't nearly as excited as he was. Only some of them agreed with his conclusion about the validity of the find. So he stored it away for 30 years.

THE FINAL VERDICT ON JAVA MAN

Before Dubois died he admitted that he had found two truly human skulls at nearby Wadjak in roughly the same level as his other fragments. Why didn't he bring these out earlier? Could it be that he felt they would distract the acclaim he hoped for in being the discoverer of a genuine missing link?

The real clincher is that before his death, Dubois declared that Java man was nothing more than a giant gibbon. It had nothing whatsoever to do with human ancestors.

And so another of the great "missing links" in man's ancestry is still mysteriously missing.

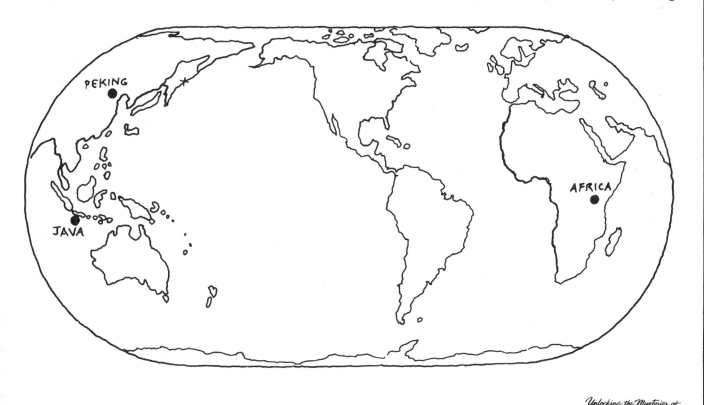

Unlocking the Mysteries of
CREATION

Are Cavemen Our Ancestors?

Before we get to modern man on the far right side of the chart we encounter two other individuals which are curiously labeled "men." This is appropriate since both Cro-Magnon Man and Neanderthal Man are both true humans indeed.

Think! If two of our evolutionary ancestors are already evolved into man, then why are they in a chart of missing links designed to show us the creatures leading up to Homo sapien?

CRO-MAGNON MAN

In 1940 some boys were out running with their dog in the countryside near Lascaux, France. The dog fell into a crack in the ground. When the boys rescued their pet they prodded their way into an ancient cavern. It was several hundred feet long and the walls were covered with colorful paintings of horses, deer, and bison.

These paintings are now famous as the skilled artwork of people we call Cro-Magnon (KRO-MAN-YO). Some of their skeletons were found buried in another cave at Les Eyzies, France in 1868. **The name Cro-Magnon simply refers to the local name of the stone cave** in which they were

found. It literally means "great big." There are more than 70 sites of Cro-Magnon art in France alone.

Based on evolutionary assumptions, the Cro-Magnon people are supposed to date back 12,000 to 30,000 years. The fact they lived in caves does not mean they were less human. Do some humans live in caves today? They do, but does that make them any less human? Realizing that, you can understand how easily tribal groups can become isolated over time and actually "de-volve" to social and technological degenerates. Is it any wonder why *Smithsonian magazine* (October 1986) carried an article titled: *"Cro-Magnon hunters were really us, working out strategies for survival".*

NEANDERTHAL MAN

The very name, Neanderthal, seems to automatically arouse thoughts of a hunchbacked primitive brute with a heavily overhanging forehead and a gorilla-like face. But what is the real story on Neanderthal Man?

The name comes from the Neander Valley near Dusseldorf, Germany. It was here in 1856 that the first skeleton of Neanderthal Man was discovered. Since then there have been many Neanderthal graves found in Europe and the Middle East.

At the time of the initial discovery and for many years after, it was publicly implied that Neanderthal Man was the missing link in man's heritage, connecting him to apes. During the late nineteenth century, with Darwin's theory shaking the scientific world, these early "ape-men" were "proof" that human evolution was a fact.

Models of Neanderthal Man were once exhibited as bent over, club-swinging cave men. But eventually it was discovered that Neanderthal men walked upright after all. In a news article of the Sacramento Union (Cali-

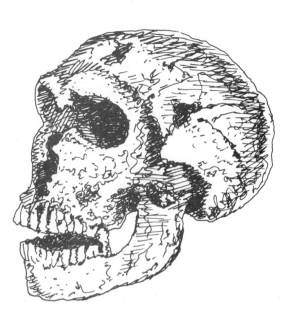

fornia) dated September 16, 1981, the sub-headline reads: **"He may not have been the hairy ape we thought he was."**

But why were Neanderthals depicted as hunchbacked and rather retarded looking? It turns out the reason is that **"one skeletal find"** proved to **"have been severely deformed by age and arthritis."**

Now the truth is known. If you were to give Mr. Neanderthal a shave and haircut, put him in a business suit, and send him downtown to pay the bills, he wouldn't stand out from the crowd at all. In fact you've likely seen individuals on the street that looked a whole lot more primitive than Mr. N.

The really surprising thing about Mr. N that few people are ever told is that **his brain size was larger than the average brain of modern man.** That seems to be evolution in reverse!

It's clear now that these people of ancient Europe were truly human in every way. Even some of their social complexity can be learned from their burials.

Is Neanderthal an evolutionary link to modern man? Not at all. An early citizen in humankind's population of the planet—certainly; but definitely not a proof of human evolution. Yes, this missing link is also still missing.

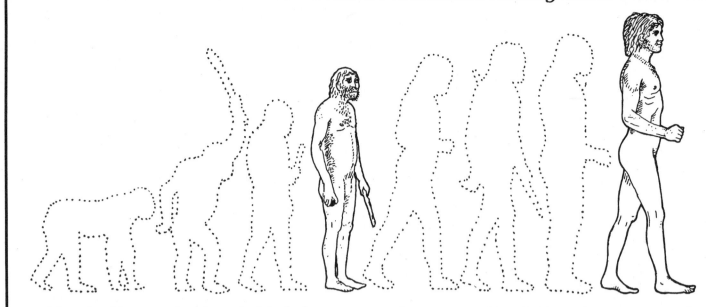

After looking at the facts behind the so-called missing links it helps us realize that the chart needs some major revision if it's going to reflect reality.

With all our adjustments, we are left with "1470 Man" who is considered by some to be almost three million years old and essentially like us. And then there's modern man walking off the end of the chart.

Where are all the other "established missing links?"

They're still missing!

There have been other famous discoveries on the chart in the past. But they have also been removed. Let's examine a couple of them.

PILTDOWN MAN

Piltdown Man was announced to the world in 1912 as the discovery of the missing link in man's evolution. The director of the Natural History Museum of London, Arthur Woodward, publicly declared this monumental discovery by a physician named Charles Dawson.

The wonderful discovery consisted of a man-like skull cap and an ape- like jawbone, unearthed at a gravel pit near Piltdown, England.

It was called "Dawn Man" and proclaimed to be 500,000 years old. The scientific com-munity was convinced. Now there was proof there really was a creature in transition between ape and man.

It was 41 years later, in 1953, when scientists finally got around to a critical analysis of Piltdown Man. They discovered that **the teeth had been filed** to fit and **the bones had been stained** to make them appear old. **The whole thing was a fraud! A fake!**

We can learn a valuable lesson here. This whole episode points out that scientists aren't infallible, nor are they always trustworthy. Like a lot of people, they tend to find what they want to find whether it's genuine or not.

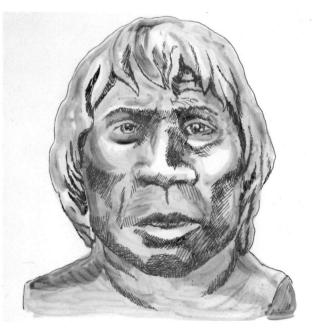

NEBRASKA MAN

The search for "the missing link" was a serious business to a number of people even in the early part of this century. Another find came from western Nebraska in 1922.

One of the most well-known anthropologists of that day was Henry Fairfield Osborn. He and some others revealed a discovery which they felt represented a creature with traits of chimpanzee, Pithecanthropus (an extinct ape), and man.

The "ape man" was given the scientific title, Hesperopithecus haroldcookii, but it came to be known to the general public as Nebraska Man. Imaginative drawings of Nebraska Man were published in the Illustrated London News on June 24, 1922. The impression given by the artwork was clearly that a primitive prehistoric caveman had been found.

When the famous "Monkey Trial" was staged in 1925 to force the teaching of evolution in American schools, the recent discovery of Nebraska Man was conveniently used as proof that man most certainly descended from lower forms of animals.

The amazing thing about Nebraska Man is that **the discovery consisted of no more than a single tooth.** But that's not all. When the truth was made known in 1927 it turned out that the tooth was actually from an extinct pig!

Unfortunately, as far as the Scopes Trial was concerned, the die had already been cast; evolution began to find acceptance in the public school classroom as if it were more scientifically valid than creation. But now we know the deception and realize that respected scientists accepted this evidence as proof of "a link."

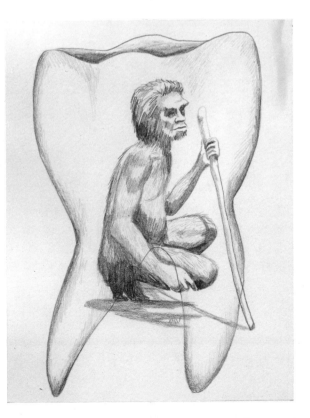

Mark Twain wrote a rather revealing insight about such cunning professional deception: "There is something fascinating about science. One gets such wholesale returns of conjectures out of such a trifling investment of facts." (Life On The Mississippi, p. 156)

Is there any evidence to link man to ape-like creatures? The cover story of *Time* magazine for November 7, 1977 featured Richard Leakey beside a naked black human. The black man is wearing a rather imaginative mask showing how "1470 Man" supposedly looked.

Candidly, the writer of the article for *Time* makes this honestly revealing statement:

"Still doubts about the sequence about man's emergence remain. Scientists concede that **their most cherished theories are based on embarrassingly few fossil fragments** and that **huge gaps exist in the fossil record.**"

It has been observed that taken all together, if you could gather all the fragments of skulls and other bones from all the so-called hominid relics found in the last century the sum total of them would not fill even one coffin!

SO WHERE IS THE OLDEST TRUE MAN?

An interview with Richard Leakey was released March 19, 1982 in the *Vancouver Sun*. According to Leakey, man's ancestors go back 3.75 million years to fossil footprints discovered by his mother, Mary Leakey, in Laetoli, Kenya.

This stunning conclusion relies of course on the doubtful ages determined by radiometric methods. But the surprise is that this discovery pushes man's earliest supposed man-like ancestors back before more primitive creatures which were once thought to be our forefathers.

The article quotes Leakey as saying: "The whole basis on which palentologists classify fossil apes and humans is misleading. The time has come to admit that the system by which we name things is inadequate in dealing with things that have a time dimension."

The latest finds Leakey refers to come from East Africa as so many of the discoveries have in recent years. *Footprints In The Ashes Of Time* was the title of the article featuring these tracks in the April 1979 issue of *National Geographic* magazine.

The volcanic ash in which the prints are found has been dated by the potassium argon method. Keep in mind the implications of that as described earlier. [see section one p. 46-47]

The tracks of many animals were also found here, "frozen" as it were in cement-like mud during a volcanic disaster.

The human prints are exactly like ones you might make the next time you walk bare-foot along a lakeshore. What do they tell us?

A variety of modern animals were associated with the animal tracks identified by expert trackers. A talented artist was commissioned to recreate the scene for the magazine. It's interesting to notice the modern appearance which the artist has given these animals. The guinea fowl in the painting are like those living today. The giraffes are also modern. The elephants are rendered with a fully modern appearance, just as are the ostrich and the hare.

But when you come to the tracks of the man how has the artist portrayed the individual who made them?

The feet look reasonably man-like. But as you look higher up the figure you get the distinct impression that the artist knows something we don't know. In these distinctly human tracks the artist has placed an ape man—some kind of half and half creature that no one has ever seen. Now you would think they could figure out what kind of creature makes human footprints. But when you see the volcano erupting in the distance and the ape-men looking the other way you wonder if the artist really has a track on reality.

Though no one has ever found any creature except human, able to make a human footprint, such a fact is irrelevant to this article. It is very clear that artists have done as much to formulate the public view of man's evolution as any scientist ever could.

A NEW THEORY OF EVOLUTION

Newsweek magazine carried an article on March 29, 1982 with a new twist on human evolutionary theory. Since the fossil finds are all either apes or men, something has to be done to explain this mystery.

"Instead of changing gradually as one generation shades into the next, evolution as (one scientist named) Gould sees it, proceeds in discrete leaps. According to the theory of *punctuated equilibrium* there are no transitional forms between species, and thus *no missing links!*"

It was once said that evolution happened so slowly that no natural examples could be found to prove it. Now some say that it happened in such quick "leaps" that no fossil evidence (links) can be found to prove it.

The apostle Paul, in his letter to the Roman Christians writes:

"They knew God but they didn't thank Him . . . Their thinking became futile and their foolish heart was darkened . . . Though claiming to be wise they became fools . . . and exchanged the truth of God for a lie."

WHERE DOES THE REAL EVIDENCE LEAD?

Unlocking the Mysteries of
CREATION

There is no evidence **in the real world** to even suggest, let alone prove, that man has evolved from some lower kind of animal!

All fossils that have ever been found are either all ape or all man. No real fossil has been proven to be transitional. In other words, **THERE ARE NO MISSING LINKS!**

The truly wise man will say with the psalmist, David:

"I will give thanks to Thee, for I am fearfully and wonderfully made; Wonderful are Thy works, and my soul knows it very well."

Psalm 139:14

Contrary to evolutionary thinking, man was made completely human from the beginning. He was a perfect specimen of human complexity. He needed nothing more to be fully functional.

WHAT MAKES MAN DIFFERENT FROM ALL THE OTHER CREATURES GOD MADE?

● It is not his anatomy, though that is unique!

● It is not his ability to walk uprightly, though that also is unique!

● It is not his ability to make and use tools!

● It is not really even his verbal speech, though that too is also uniquely man's gift!

● It is not just his great intelligence, as wonderfully superior as that is!

MAN IS UNIQUE BECAUSE OF HIS DEEPLY SPIRITUAL NATURE

● *Made in God's Image*

In Genesis 1:26 God said, *"Let us make man in our image according to our likeness..."*

What is God's likeness? God is not a man of

flesh and bone. As Jesus said in John 4:24, *"God is Spirit..."* And that is primarily the nature of man that makes him distinct from all other creations of God.

Unlike all the animals, God made man in the class of spirit beings. Yes man lives in a body of flesh, but his eternal God-made spirit makes him the highest and only creation who can operate beyond the limits of mere fleshly animals.

● *How Close To God Is Man?*

In Psalm 8:5 we read that God has made man *"a little lower than the angels..."*

The original Hebrew word in this text for "angels" is ELOHIM. It is one of the Biblical names for **God Himself!**

In What Realm Can Spirit-Man Think?

In Philippians 4:8 we are given some guidelines on how to operate our minds. Whatever things are true, honest, just, pure, lovely, good, virtuous, and praiseworthy; these are the kinds of abstract thoughts available to man. No animal is capable of thinking in these realms.

Only Man Appreciates Beauty.

In Psalm 27:4 we are told to *"Behold the beauty of the Lord."* Indeed every corner of the infinite creation is filled with the Creator's endless treasure of beauty.

When we humans go out and explore the beautiful world God made, we alone can truly draw deep inspiration from it. Our innermost feelings are touched by the breathtaking majesty of an autumn sunset. We can thrill to the crescendo of a symphonic orchestra. Even the shimmering iridescence of a tiny tropical fish can stir our sense of awe.

Show any of these things to a hog or a horse and they'll just ignore your discovery and go back to their eating.

What Is Spirit-Man's Authority?

In Genesis 1:26 God reveals His amazing intention to give man the total dominion over every other creature on the Earth. Psalm 8:6 and Hebrew 2:8 reflect this too.

No other creature in Heaven or Earth was given such supreme rulership. God has left nothing that is not subject to man.

Man was designed to be a ruler not an animal. He was created to fellowship with God, not to wander far from Him.

Not only has God made man like Himself in terms of spiritual potential, but God has made man just a little lower than the Creator. **Man is not just a little higher than the monkeys. He's just a bit below God!** Think of the implications of such a truth.

How Can Man Relate To God?

In 1 John 1:3 we read that *". . . our fellowship is with the Father and with His Son Jesus Christ."*

Unlike any of the animals, man is able to enjoy a personal and meaningful relationship with God. And it is clear from the Bible that the only way to achieve that fellowship is through the Master of Creation and Life itself, Jesus Christ.

Despite all the intellectually vain imaginations about Mankind's beginning, the clear evidence reflects the Biblical truth that Man has always been Man from the beginning.

In the beginning of man's history the Deceiver of all time managed to skillfully convince our first parents that they should listen to him and not obey God. Satan is still doing that today.

One of the most subtle and wicked deceptions Satan has used is the lie that men are getting better along with the rest of the natural world. Man's dignity has been robbed from him. As a result the enemy, Satan, has managed to convince men they should act out what they feel are natural animalistic tendencies. Not even the basest animals behave with such perversion.

God says: *"See to it that no one takes you captive through philosophy and empty deception, according to the tradition of men, according to the elementary principles of the world, rather than according to Christ."* (Colossians 2:8)

A captive is a slave. Throughout history the victorious armies became the brutal masters of the nations they spoiled. Such slavery is demeaning, depressing, and, ultimately deadly.

Is it possible that philosophies can have such painful results? Can men's traditions lead us into a deadly trap? God reveals more through the prophet Isaiah (in chapter 5, verses 12 and 13) saying:

" . . . they do not pay attention to the deeds of the LORD, nor do they consider the work of His hands. Therefore My people go into exile for their lack of knowledge." (NAS)

The Creator and true science have the same objective! To expose deception and reveal the Truth will always lead to greatest liberty. That is why God says:

"Test all things. Hold fast what is good."

"WHATEVER IS BORN OF GOD OVERCOMES THE WORLD: AND THIS IS THE VICTORY THAT HAS OVERCOME THE WORLD—OUR FAITH. AND WHO IS THE ONE WHO OVERCOMES THE WORLD, BUT HE WHO BELIEVES THAT JESUS IS THE SON OF GOD?"

1 JOHN 5:4-5

SUPPLEMENT
TO SESSION THREE

EVIDENCE OF HUMANS
BURIED BY THE FLOOD

*"THEN GOD SAID TO NOAH,
'THE END OF ALL FLESH HAS COME
BEFORE ME;*

*FOR THE EARTH IS FILLED WITH VIO-
LENCE BECAUSE OF THEM;*

*AND BEHOLD, I AM ABOUT TO DE-
STROY THEM WITH THE EARTH.'"*

GENESIS 6:13 (NAS)

Why Have Some Ancient Human Skeletons Been Ignored?

OVERCONFIDENCE IN THEORIES CAUSES RESISTANCE TO FACTS

Have you ever heard the quip:

"Don't confuse me with the facts, my mind is already made up!"?

There is a substantial amount of evidence that clearly destroys the popular concepts of organic evolution and Earth history. However, because the well-established systems of science, education, and public media are thoroughly committed to the belief that macro-evolution is a "fact," you will seldom hear the FACTS about discoveries which totally upset the belief.

WHAT IF THERE WAS A GLOBAL FLOOD?

We will have to leave the many fascinating details of the great flood for another volume which will follow this one. But if the flood really happened as described in the Bible you would expect to find a number of key evidences on the Earth today. There should be:

● Many cultural traditions of the flood story

● Thousands of feet of flood-deposited sedimentary rock layers covering most of the Earth

● Evidence that many creatures have become extinct by the disaster

● Buried remains of all manner of life forms found mixed up in the sedimentary layers

● The discovery of so-called "higher animal" skeletons (including man) at many different depths in the flood-produced rocks

Indeed, all of these things **have been confirmed repeatedly** over the years. But when it comes to the interpretation of the rocks of the Earth the theory of "gradualism" (as described earlier) holds an upper hand. From the bottom of the Earth's rock layers, we are supposed to find a gradual progression of life forms. The so-called "simple" ones are expected to be found at the bottom. Then,

proceeding through over half a billion years of the supposed rock pages of Earth history, we are told that evidence of Man's presence will only be found near the top, in the last million or so years.

THINK! How would modern theories be affected if truly human bones or sophisticated man-made artifacts were found deep in the sedimentary rock layers of Earth's crust?

If only one such verified discovery were made, the whole gradualistic scheme of evolution would be in ruins!

Let's look at some amazing discoveries that are virtually unknown to the general public.

FROM GERMANY

In 1842 coal miners in Germany came across a very surprising fossil in the coal seam they were mining. Bringing it into the light, it was verified to be a genuine human skull! It can be seen today as part of the collection of the Freiberg Mining Academy in West Germany.[1]

Why was the skull dismissed as a fake back then? Only because it was found in a deep stratum of brown coal, estimated to be as much as 50 million years old! (according to evolutionary reckoning). More recent analysis of the skull has pointed out that the skull, in fact, was genuinely buried along with the plants that make up the coal seam in which it was found.

But Wait? How can 50 million year old coal contain remains of human kind which supposedly only go back a million years or so?

WHERE DID COAL REALLY COME FROM?

The evolutionary assumption holds that coal was formed gradually in swampy forests that lived and died at the site of the coal formation over the course of thousands of years. The coal itself is supposed to be simply the compacted accumulation of forest plants which were eventually covered by more sediments that took millions of years to stack up.

Coal deposits do not verify such a model. In fact the well-preserved remains of leaves and insects found often in coal show that the whole mass of vegetation had to have been buried very quickly as in a massive flood. The fact that coal seams are found from just a few inches thick to scores of feet in repeated overlying layers, separated often by sterile rock types with no sign of soil, also shows what would be expected by the tidal wave activity of a global deluge.

If a human skull was really found embedded naturally in a coal seam it could very well have belonged to a contemporary of Noah who lived on the Earth about 4,500 years ago.

MANY OLD BONES FROM VARIOUS PLACES

In 1958 a coal mine in Italy yielded a human jawbone from 600 feet deep, in the "Miocene stratum," dated by the evolutionist's chart at 20 million years old.[2]

During the California Gold Rush of the 1850's there were thousands of men digging beneath presumably ancient lava flows of the Sierra Nevada mountain region. Many puzzling relics were reported. The Calaveras skull is the most famous. It was found beneath repeated layers of volcanic rock presumed to be millions of years old. Yet the skull is very modern.[3]

In 1866 a complete human skull was dug out from under a layer of volcanic basalt. Experts from the State of California and Harvard College investigated and concluded the skull to be very modern (i.e. human). The "mystery" that bothered them was how the human skull and artifacts got into a rock layer 12 million years old (according to evolutionary dating).[4]

All these discoveries and many more are thoroughly documented by original sources. They can be examined further in the reports of others who have collected the information. (See references).

BONES IN STONE IN UTAH

One interesting account came out in 1975 about a new discovery of bones in the desert of Utah. A rock collector named Ottinger found some teeth he realized must be important. When he brought experts and even cameramen back to verify the find, they wound up excavating the lower halves of two human skeletons. Every indication showed these bodies were encased at the time the rock itself was actually laid down. The bones were taken to the University of Utah for official testing and confirmation. Nothing was done; no report was issued. The rock collector eventually had to go claim his bones. Why wasn't this impressive discovery of early man in America followed up? Could it be that **the fact the bones came out of a rock layer supposedly 100 million years old, scared off the experts?** Such a "mystery" would be too much for the accepted geologic chart to endure. After all no one (among evolutionists) would believe that there could have been a true human 100 million years ago.[5]

But if the geologic chart is wrong and the flood really did bury these people, then the evidence fits. No other reasonable explanation has even been suggested. So why have the facts been ignored?

Unlocking the Mysteries of **CREATION**

SUPPLEMENT
TO SESSION THREE

UNLOCKING THE MYSTERIES
OF THOSE TERRIBLE LIZARDS,
THE DINOSAURS

"GOD SAID TO THEM,
'BE FRUITFUL AND MULTIPLY; FILL
THE EARTH AND SUBDUE IT;
HAVE DOMINION OVER . . . EVERY
LIVING THING
THAT MOVES ON THE EARTH.'"

GENESIS 1:28

There probably isn't another subject more fascinating to young and old alike than the dinosaurs.

You probably know some boys about ten years old who just love to play with and even collect models of the dinosaurs.

One of the finest displays of life-like replicas of dinosaurs is in Canada. The pre-historic park at the Calgary Zoo in Alberta is a great experience with extinct giant animals.

Coming to the entrance of the park you'll notice an important sign to set the stage for your tour.

THE DINOSAURS REIGNED FOR 140 MILLION YEARS. THEY MYSTERIOUSLY BECAME EXTINCT ABOUT 60 MILLION YEARS BEFORE MAN APPEARED ON EARTH.

This concept is very typical. When we think of the dinosaurs, most of us automatically relate to "millions of years ago," long before man ever arrived on the scene.

Great lumbering beasts like the Triceratops are nowhere to be found walking the Earth today.

TRICERATOPS (try-SAIR-uh-tops)

● Name means "3-spiked head."

● Overall length up to 25 feet (about as long as a large delivery truck!)

● Up to 10 feet high at the ridge of his back.

● Weighed up to 24,000 pounds.

● From horned collar to nose, the head was about 8 feet long.

● Two massive horns over the eyes, 40 inches long and almost a foot wide at the base.

● A strikingly handsome and powerful creature that thrived in great numbers before the flood.

TYRANNOSAURUS (tie-ran-uh-SAWR-us)
Name means "tyrant lizard"

The terrible Tyrannosaurus, we're told, is supposed to have suddenly died out some 60 million years ago.

● Up to 50 feet long (as long as a railroad boxcar).

● As high as 18 feet, he could rest his chin on the roof peak of an average house.

● Weighed up to 20,000 pounds. Skull length measured over 4 feet.

● Claws on hind feet up to 8 inches in length.

● Teeth like daggers up to 6 inches in length.

● Small forelimbs seem mysteriously useless to the experts.

● Many feel this was probably the fiercest animal that ever left its tracks on Earth.

The Mysterious Vanished Giants

A Variety of strange giant creatures once flourished all over the Earth. We're told that the "Age of Reptiles" lasted over 100 million years. Then they all mysteriously vanished into extinction.

All we find now are the bones of these once-great Titans that make present day animals appear like dwarfs in contrast. Men dig their petrified skeletons out of the layers of the Earth. Then they are carefully prepared and reassembled for display in museums all around the country.

DIMETRODON (die-MEE-tro-don)

● Unusual for its sail-like fin, the purpose of which is a mystery to the experts.

● Up to 11 feet long.

● Weighed over 650 pounds.

SCOLOSAURUS (sko-luh-SAWR-us)

● This low-to-the-ground "living tank" must have been an invincible vegetarian with all his armor.

● Up to 18 feet long and 8 feet across the midsection.

● Covered with spike-studded armor, the knobs stuck out 4 to 6 inches.

● His bony-knobbed tail wielded two spikes to ward off unwelcome antagonists.

STEGOSAURUS (stegg-uh-SAWR-us)

● Famous for its "second brain" located along the spine above the hips.

● Up to 25 feet long.

● Huge armor plates along its spine have mystified the experts.

● Up to 12 feet high at the rear legs.

● Curiously built with low front legs and head low to the ground.

● Could weigh up to 20,000 pounds.

GLYPTODON (GLIP-tuh-don)

● Another "living tank" as big as a rhinocerus or even up to 15 feet long.

● Not a reptile, but a mammal giant of the past that resembled the armadillo living today.

● The bony outer casing was like a series of overlapping scale-rings, enabling it to bend its body.

● Spikes on the knob-end of its tail make it look especially suited for battle against any creature daring enough to think he could get close to it.

What Extinct Animals Lived With Man In The Past?

POST ICE AGE MAN

Question: Do you suppose these great creatures were ever seen alive by human beings?

Consider the implications of such a question, especially now that you know what you do about the Biblical view of Earth's history.

4,000 Years Ago

The ancient record of a man named Job is found in the Bible. It goes back to a time over 2,000 years before Christ.

The fascinating story of Job is one of the oldest pieces of literature on Earth. It was written down just a few hundred years after the flood of Noah's time.

In the 40th chapter of Job's account we see the record of the Creator Himself speaking to Job. He is drawing Job's attention to one of the wonders of the creation. Let's read it.

"Look now at the behemoth, which I made along with you . . . " (verse 15)

God is saying to Job, "take a good look at this creature. I'm going to demonstrate something for you."

Some Bibles have a marginal note by this verse, identifying this animal as an elephant or hippopotamus. Well, let's investigate further. Is that really what Job is looking at?

"He eats grass like an ox." (verse 15)

The animal must be a vegetarian, but it's likely larger than an ox.

"See now, his strength is in his hips (or loins)." (verse 16)

This critter must have powerful legs. It could be an elephant.

Elephant

Hippopotamus

"His power is in his stomach muscles." (verse 16)

He apparently has a massive mid section. If that was all we had to go on we could assume Job may very well be looking at a hippo. But there's more information yet.

"He moves his tail like a cedar." (verse 17)

Now we've got a problem. Have you noticed something distinctive about cedar trees? They are big aren't they? Now have you ever seen the tail of an elephant or a hippo? It doesn't seem that a cedar tree is a very appropriate analogy does it? But what other large animal do we know of that lived on the Earth with Job?

Giant Sloth

Mammoth

Ice Age Giant Animals

Look at the tail of the Brontosaurus. We know that giants like these really did roam on our Earth. Can you imagine the mighty demonstration of force as he moved along. He truly is a good example of the awesome power of God displayed in the animal kingdom isn't he?

And just what does it go on to say in the 40th chapter of Job about this beast?

"HE IS THE CHIEF OF THE WAYS OF GOD!"

"His bones are like beams of bronze. his ribs like bars of iron." (verse 18)

"He is the first (or chief) of the ways of God." (verse 19)

Whatever this creature is, it's described as a supreme demonstration of God's power. Could it possibly be a DINOSAUR?

Next question: Is there any **evidence** that the gigantic animals of the past were ever seen alive by humans?

"BEHOLD NOW BEHEMOTH!"

In the Paluxy River bed of Texas (near the town of Glenrose) there are layers of hard limestone rock that have attracted attention for most of this century. Why?

It just so happens that many petrified foot tracks have been found there, made by a variety of dinosaurs. Supposedly, these tracks were made about a hundred million years ago, when creatures walked across flats of cement-like mud.

A state park has been established to preserve these unique clues to the dim past. There are many tracks here that visitors can see clearly. The surrounding areas also have tracks.

These tracks were made by creatures of immense size, none other than the gigantic lizards of the past. You can see how well-preserved they are, even though the river has been washing over them year in and year out. Even that makes you wonder why they haven't eroded away in all those supposed millions of years doesn't it?

Recently there has been renewed interest in this area. Research teams have been scouring the bedrock of the river. The reason?

Reports of local residents have claimed for years that there are other tracks here at the Paluxy, made by humans!

Unlocking the Mysteries of CREATION

Did Man And Dinosaur Live Together?

The Paluxy River in Texas holds evidence that has aroused substantial controversy in recent years. Petrified footprints of what appear to be humans have been reported in the same rock alongside the tracks of behemoth animals like the mighty brontosaurus.

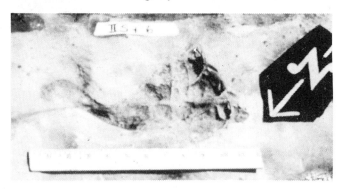

WHAT COULD HAVE MADE THESE TRACKS?

For years the skeptics passed these mysterious tracks off as the creative work of Indian carvers of long ago.

Why would they say that? They never suggested such an idea about the dinosaur footprints?

The reason some people felt there was a problem here is because of the traditionally accepted geologic chart. According to it, the dinosaurs supposedly became extinct 60 million years before man evolved.

Where did that idea come from? It came from the imaginations of men like Hutton, Lyle, Kant, and Darwin. These evolutionists denied the reality of the Biblical account of creation by God. They disregarded the historical and scientific evidence for a world-wide flood.

The standard geologic chart is a theory!

In 1954 the *National Geographic* magazine ran an article titled, "We captured a live Brontosaur." They were highlighting these same Texas-size footprints from the Paluxy River.

They sand-banked the river to uncover and remove a whole line of consecutive Brontosaurus footprints. Eventually the well-preserved tracks were installed in the Brontosaurus exhibit at the American Natural History Museum in New York City. But nothing was said about the other mysterious tracks found there!

Why wasn't the problem at least acknowledged?

Could the human looking tracks found at the Paluxy River really have been made by humans walking on what was soft cement-type mud at the same time the dinosaurs made their tracks here? The controversy became a major issue during the 1980's, but there have been some erroneous reports circulated which have caused considerable confusion on the topic.

After having been to the site several times and spoken to a number of scientists, both evolutionists and creationists, we can safely observe several important conclusions.

The river has exposed numerous obvious dinosaur tracks ranging in size from less than a foot long to some that are more than two feet across. The tracks are petrified; that is they were made in a kind of cement sediment that is not typically being formed today. Many trails of consecutive left-right-left tracks are clearly seen in the flat river bottom.

EVIDENCE THAT THESE TRACKS MAY VERY WELL HAVE BEEN MADE BY HUMAN BEINGS WHO LIVED CONTEMPORARY WITH THE DINOSAURS. This was not millions of years ago, but just several thousand years ago at the time of the world-wide flood and other subsequent catastrophes which are responsible for the deposition of our planet's layers and land forms. But that's not all.

Beside the possibility of man tracks in supposedly prehistoric strata from some evolutionary era of dinosaurs, there are other things as well. Dinosaur skeletons found where evolutionists didn't expect them; human artifacts in layers even deeper; supposedly more ancient fossils (like trilobites) in surface rocks which evolutionists claimed were millions of years younger, and more.

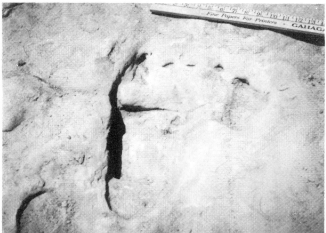

The human looking tracks do indeed have distinctively human characteristics like the big toe, ball of the foot, narrower arch, and rounded heel. Being more delicate than the dinosaur tracks, human tracks would be expected to erode and become less distinct even after a few months of exposure. Over several decades more and more human looking tracks have been exposed by the removal of heavy slabs of original sedimentary "overburden." More than a hundred such tracks have been documented. Some are found in sequence with others that are identifiably human in appearance.

In 1986 news stories circulated that creation scientists had finally given up. A man had found curious rust stains indicating dinosaurian claw marks at the front of one particular series of tracks. The stains are there indeed. But they are only associated with a single trail of tracks, several of which show evidence of human foot tracks in the middle of the dinosaur tracks.

The plot thickens. More research continues to be done. The sweeping claims by evolution-biased media that the man-track evidence is debunked is just plain misleading! THERE IS SUBSTANTIAL

There are some puzzling mysteries at the Paluxy River. Some fossil tracks appear to have been made by large mammals like bears and cats. Even claw marks are preserved. **The question is: "How did these animals (which are supposed to be a much later evolutionary development) step back millions of years to the 'time of the dinosaurs'?" Or did they both live together in much more recent time, even since the great flood?**

Other petrified ancient human footprints have been found in a variety of places. A series of clear human prints were found in hardened volcanic ash in Nicaragua. We're told they are 2,000 to 5,000 years old. It's a good thing they didn't find any pteradactyl tracks nearby or we probably would never have heard about the human tracks.

Keep in mind that the traditional geologic chart is only imaginary. The millions-of-years gap between man and the dinosaurs is not proven by facts. **The facts of other "out of place fossils"** (as previously mentioned) **should alert us to the problems with the theory.**

THE SOURCE OF DRAGON LEGENDS

Why is it that ancient civilizations around the globe hand down traditions about dragons universally?

Of course many of the stories are admittedly fanciful. But do myths of ancient peoples have some basis in fact?

Let's read from that 4,000 year old account about Job again. In the forty-first chapter, God is talking to Job and asking some rhetorical questions.

"Can you draw out Leviathan with a hook?" (verse 1)

The implied answer is "of course not . . . this animal is famous for its immensity and fierceness." But let's read on.

7 "Can you fill his skin with harpoons, or his head with fishing spears?

8 "Lay your hand on him; remember the battle—never do it again!

9 "Indeed, any hope of overcoming him is in vain; shall one not be overwhelmed at the sight of him?

15 "His rows of scales are his pride shut up tightly as with a seal;

17 "they are joined one to another, they stick together and cannot be parted.

18 "His sneezings flash forth light, and his eyes are like the eyelids of the morning.

19 "Out of his mouth go burning lights; sparks of fire shoot out.

20 "Smoke goes out of his nostrils, as from a boiling pot and burning rushes.

21 "His breath kindles coals, and a flame goes out of his mouth.

22 "Strength dwells in his neck . . .

25 "When he raises himself up, the mighty are afraid; because of his crashings they are beside themselves.

from a nineteenth century engraving

27 "He regards iron as straw, and bronze as rotten wood.

30 "His undersides are like sharp potsherds; he spreads pointed marks in the mire.

31 "He makes the deep boil like a pot; he makes the sea like a pot of ointment.

32 "He leaves a shining wake behind him: one would think the deep had white hair.

33 "On Earth there is nothing like him, which is made without fear.

34 "He beholds every high thing; he is king over all the children of pride."

To our sophisticated ear that may sound fanciful and ridiculous. Imagine, a fire-breathing monster too fierce for man's usual hunting methods to subdue it.

Unbelievable?

Just because we haven't seen it does that mean that it never existed?

What about some of the other strange creatures in our present world?

If it weren't for the fact that fireflies were so commonplace, we would tend to be skeptical about living, flying lightbulbs.

Little beetles shooting explosions at their enemies also seems incredible.

And the idea of a creature actually producing an electric shock, as does the eel, would indeed be unbelievable if it weren't a proven fact.

Some of the strange breathing passages of dinosaur skulls gives cause to wonder if they had special features such as Job saw.

A fire breathing dinosaur is entirely within the realm of biological possibility!

Could Some Dinosaurs Still Be Living?

WHAT IS IT?

In the spring of 1977 off the coast of New Zealand, a Japanese fishing crew hauled in a very peculiar "catch" in their nets.

The **gigantic creature was 30 feet long** with a husky neck about five feet long! Its bulky body **weighed about 4,000 pounds.**[6]

The "monster" had been dead for quite some time. In such decayed condition it would have spoiled the valuable cargo of fish had it been kept aboard. So the fishermen photographed it and studied it as best they could before throwing it overboard into the ocean. Others reported having seen these creatures alive in the nearby waters.

It definitely looks like a reptile or amphibian, rather than a fish. The animal appears

remarkably similar to the dinosaur known as the plesiosaur. But the plesiosaur, we're told, became extinct millions of years ago.

This is a baffling mystery to the doctrine of evolution. But from the Biblical prespective there is no reason why some of these creatures could not have survived in the depths of the sea since the great flood.

IS THERE REALLY A LOCH NESS MONSTER?

In the Highlands of northern Scotland people have been reporting a gigantic swimming "beastie" for the last 1,400 years.

Some **3,000 sightings** of the creature have been tallied. Pictures have been taken of it which **look remarkably like a plesiosaur.** It's said to be about 20 feet long with a serpent-like neck and fat body with flippers on the sides.

Because of the extreme depth of the Loch (lake) Ness, many people believe the amphibious "Nessie" may actually be more than one giant sea serpent residing in some dark submarine cavern.

Could a few of the world's giant sea monsters have survived the flood to produce descendants reaching 5,000 years into our day?

DRAGON LIZARDS OF KOMODO

Imagine taking an afternoon stroll in the woods and bumping face-to-face into a ten-foot-long dragon!

The world's largest living lizard, according to a *National Geographic* article of December 1968, the Komodo monitor lizard "remains something of a mystery."

There are about a thousand of these terrifying lizards surviving on the remote island of Komodo in the Indian Ocean north of Australia.

They are called living examples of the prehistoric Age of Reptiles. They were **totally unknown to modern man until the year 1912!** Could other "dinosaurs" still be living today?

DINOSAURS IN THE CONGO?

Deep in the heart of tropical Africa, the native pygmy people tell of strange encounters with giant dinosaur-like reptiles in modern times.

Occasional visitors to an unexplored part of the Congo have returned with incredible stories over the last two centuries. They all confirm that the natives are absolutely honest in their reports of the "mokele-mbembe" (mo-KEL- ly-mm-BEM-be). They say these animals are huge, with small heads, long necks, and long massive tails, they wade in slow meandering rivers, and have been described as **"half elephant and half dragon."**[7]

Interviews with the natives have revealed astonishing things. About 20 years ago some of them managed to spear one of the creatures to death. All those who ate it, died soon afterward. When shown pictures of various large animals, living and extinct, the pygmy people there always verify the brontosaurus is most like what they call mokele-mbembe!

Just think what the discovery of a living Brontosaurus would do to the mythical evolutionary chart of geologic history?

Unlocking the Mysteries of
CREATION

Why Did Dinosaurs Become Extinct?

INADEQUATE THEORIES LEAVE ONLY UN-SOLVED MYSTERIES

One of the greatest mysteries of Earth science over the years has been the extinction of the dinosaurs. The traditional non-Biblical concept holds stubbornly to the idea that the Earth was dominated by the giant reptiles for over 100 million years. Reading the popular literature one gets the idea that mammals couldn't really get a foothold on Earth until the giant reptiles were out of the way. Of course the evolution myth insists that humans finally became the dominators only after another 60 million years of transmutational struggle for rulership.

A recent issue of *Time* magazine featured a cover story about a new theory to explain the mysterious extinction of the dinosaurs (5/6/85).

Over the years science theorists have come up with a variety of bizarre explanations. Every one of them has been discredited by careful logic and investigation. Let's look at some of them.

Some people thought that "early" **mammal rodents** managed to eat all the dinosaur eggs. But the suddenness of dinosaur demise insists this theory can't work, especially in view of the fact that such mammals supposedly lived with the giant reptiles for millions of years.

Another idea supposes that the **brains** of the dinosaur kind eventually made them obsolete. But they seemed to manage all right with those brains from the beginning of their existence.

Could **overweight** have been the problem? Some think their tremendous bulk eventually caused them to get "slipped disks" in their spines, thus incapacitating them from foraging for food.

In 1946 one palentologist suggested a **world-wide heat wave** of two degrees above normal would have "baked" all the testicles of the male dinosaurs. With mass sterility setting in, it wouldn't be long before every last dinosaur died without heirs.

Perhaps they were **fussy eaters** and their routine staple plantfood suddenly became unavailable: a sure case of starvation. Others suppose the dinosaurs didn't have enough sense to avoid **poisonous plants** and all of them succumbed due to their lack of dietary discretion.

Believe it or not, one theory suggests, in all sincerity, that the dinosaurs may very well have died out because of a mass case of **constipation!**

The cover of *Time* asks: *"Did Comets Kill The Dinosaurs?"* Proponents of a new theory are emphasizing revolutionary evidence which they feel reflects a series of **cosmic catastrophes** on Earth.

A tremendous impact supposedly raised so much dust that the sky darkened, the world temperatures dropped, and the reptiles died of frostbite. As *Time* reported, "Whether **these catastrophic impacts** are random or cyclic remains to be seen. But if they occur at all, they **could shake the foundations of evolutionary biology . . . "**

The implications do present a major break with long-held evolutionary theory. But what they are still missing are the phenomenal implications of a globe-rocking cosmic collision.

Nevertheless, the new catastrophe theory does face facts that other extinction models ignore. As *Time* put it, "all these fanciful concepts fail to account for the hundreds of other species that perished at the end of the Cretaceous." U.C. Berkeley physicist, Luis Alverez, says *"The problem is not what killed the dinosaurs but what killed almost all the life at the time."*

In skillful, though misleading, editorial style *Time* concludes the article by quoting another scientist, saying: "Maybe there just wasn't enough room for them on the ark."

"Always learning and never able to come to the
knowledge of the truth."
2 Timothy 3:7

THE CHOICE IS OURS TO DISCOVER
THE TRUTH
IF WE ARE TRULY WILLING.

"Just as my mouth can taste good food,
so my mind tastes truth when I hear it."
Job 12:11 (TLB)

SUPPLEMENT
TO SESSION THREE

THE BRILLIANT GLORY OF
EARTH'S EARLIEST KING

*"Those who are wise will shine
like the brightness of the heavens,
and those who lead many to righteousness,
like the stars for ever and ever."*
Daniel 12:3 (NIV)

The Bible gives us wonderful insights if we'll take the time to study and pray. As it says in Proverbs 3:5, we must put our confidence in the LORD and not in our own shallow understanding. As we do that, he will guide us into all the truth.

The following study outline will bring out some little-known facts that are mentioned in the Bible. As you conclude this section about Unlocking The Mysteries Of Original Man, may "the God of our Lord Jesus Christ, the Father of glory, give to you a spirit of wisdom and revelation in the knowledge of Him." (Ephesians 1:17)

What happened to Adam and Eve that caused them to "see" their nakedness immediately after they rebelled against God's spoken will?

Just how brilliantly beautiful was the man without imperfection, who was created in the very likeness of his Maker?

• • •

WHAT PART DOES LIGHT PLAY IN RELATIONSHIP TO THE PRESENCE OF GOD?

● Jesus

"And while He was praying, the appearance of His face became different and His clothing became white and gleaming (literally flashing like lightning)." Luke 9:29

"I am the light of the world."

John 8:12

● Moses

"Moses did not know that the skin of his face shone because of his speaking with Him . . . the skin of his face shone and they were afraid to come near him (vs. 30) . . . the skin of Moses' face shone. So Moses would replace the veil." (vs. 35). Exodus 34:29ff

● God

"God is light, and in Him there is no darkness at all." 1 John 1:5

"Thou (my God) art clothed with splendor and majesty, covering Thyself with light as with a cloak . . . " Psalm 104:1-2

" . . . the King of kings and Lord of lords; who alone possesses immortality and dwells in unapproachable light."

1 Timothy 6:15-16

"And now men do not see the light which is bright in the skies; . . . out of the north comes golden splendor; around God is awsome majesty."

Job 37:21-22

"How blessed are the people who know the joyful sound! O Lord, they walk in the light of Thy contenance."

Psalm 89:15

Isn't it interesting how the inspired word of God frequently associates the presence of God Himself with awesome brillant light? Think how physical light is so divinely unique. It is incomprehensible. Science can not adequately come to terms with light. Its very essence is divine. The starry lights of the heavens declare the very glory of God with subtle, yet profound symbolism.

When God spoke at the very beginning, there was light. Down through time man has been spiritually affected by real physical light. Is there a reason for it?

How will the people of God appear in the future when with Him?

"And there shall no longer be any night; and they shall not have need of the light of a lamp nor the light of the sun, because the LORD God shall illumine them; and they shall reign forever and ever," (compare Rev. 21:23 "...the city...the glory of God has illumined it.")

Revelation 22:5

"...the Lord will be to you an everlasting light, and your God your glory...for the Lord will be your everlasting light..."

Isaiah 60:19-20

"the glory of the Lord shall be revealed, and all flesh shall see it together..."

Isaiah 40:5

"For the Earth will be filled with the knowledge of the glory of the Lord, as the waters cover the sea."

Habakkuk 2:14

"Those who are wise shall shine like the brightness of the firmament, and those who turn many to righteousness like the stars forever and ever."

Daniel 12:3

"Beloved, now we are children of God; and it has not yet been revealed what we shall be, but we know that when He is revealed, we shall be like Him, for we shall see Him as He is."

1 John 3:2

"When Christ who is our life appears, then you also will appear with Him in glory."

Colossians 3:4

"giving thanks to the Father who has qualified us to be partakers of the inheritance of the saints in the light."

Colossians 1:12

There is no way for us to fully appreciate the quality of life Adam and Eve enjoyed before their sin. Neither is there any way for us to really grasp the excellent relationship they had with their loving Creator. We know they had a close fellowship with God in person. Their innocence may very well have been bathed in God's glory. Could that glorious light have been dimmed when sin broke their dynamic connection with Spiritual Life Himself?

"And they were both naked, the man and his wife, and were not ashamed."

Genesis 2:25

"Then the eyes of both of them were opened, and they knew that they were naked..."

Genesis 3:7

If Adam's "light" was suddenly turned out, that may give some insight to a new testament reality.

"But if we walk in the light as He is in the light, we have fellowship with one another, and the blood of Jesus Christ His Son cleanses us from all sin."

1 John 1:7

God's light is now revealed through Jesus Christ the Redeemer, and through His life-giving Word. To Adam the glory of redemption was not known, but he did experience a brillant glory of God's manifest presence continually.

FROM GLORY TO GLORY

2 Corinthians 3:18

"But we all (believers), with unveiled face, beholding as in a mirror the glory of the Lord, are being transformed into the same image from glory to glory, just as by the Spirit of the Lord." [Don't confuse the meaning of this phrase with Romans 1:17, speaking of spiritual growth "from faith to faith."]

Notice the parallels of what we have seen so far.

1. God is clothed in radiant glory.

2. The two men on Earth who experienced the most transcendant manifestation of God's presence literally shone like lights because of the exposure to the heavenly glory.

3. God's saints in the future will "shine like lights" because of their intimate presence with God.

Do you think it possible that the original glory of Adam in the garden may have given him the appearance of being clothed in light? Could it be that their physical nakedness was insignificant compared to the brilliance of their spiritual likeness to God?

That phrase in 2 Corinthians 3:18 says we are being transformed into the same image. What is that image? Was not man made in God's image? He lost that glory because of sin. He became "dead in sin." The glory that original man had was wonderful. But the glory to come, through our merciful Redeemer, Jesus Christ, is far more glorious!

THE DESTINY OF GOD'S CHOSEN PEOPLE

1 Peter 2:9

"But you are a chosen generation, a royal priesthood, a holy nation, His own special people, that you may proclaim the praises of Him who called you out of darkness into His marvelous light."

When you don't know God you are in spiritual darkness. If you don't take that first step of faith in the darkness toward Jesus, then you just keep groping in the confusion of hopeless darkness. Here is where the Living Word comes in and breaks up the deception of that seemingly endless darkness.

THE GLORIOUS HOPE

2 Corinthians 4:3

"But even if our gospel is veiled, it is veiled to those who are perishing." (the context is continuing from chapter 3, relating the experience of Moses covering his glowing face with a veil).

verse 4

The perishing are those "whose minds the god of this age has blinded, who do not believe, lest the light of the gospel of the glory of Christ, who is the image of God, should shine on them."

verse 6

"For it is the God who commanded light to shine out of darkness (note the context is creation) who has shone in our hearts to give the light of the knowledge of the glory of God in the face of Jesus Christ."

Where was the glory of God seen by the disciples on the Mount of Transfiguration? In the face of Jesus. Through whom can we be enlightened with the knowledge of God's glory? Again, it's Jesus.

verse 7

"But we have this treasure in earthen vessels, that the excellence (surpassing greatness) of the power may be of God and not of us."

What is the treasure? The knowledge of God. Where is the treasure? In a very unglorious earthen vessel of flesh. That glory is there inside the believer in Jesus, but it has yet to be revealed and shine out. An Old Testament story tells us about other earthen vessels to beautifully illustrate the impact of this truth.

A DARING ADVENTURE OF FAITH

The Apostle Paul recognized a wonderful spiritual truth when he wrote the fourth verse of Romans chapter 15. "Whatever things were written before (the Old Testament) were written for our learning, that we through the patience and comfort of the Scriptures might have hope."

Just following the life of a man like Gideon in the Old Testament book of Judges is an encouragement to anybody who feels there is no hope. But one particular event in his life demonstrates a God-given symbolism worth noticing.

In Judges 7:16 we see Gideon leading his 300 brave and obedient men against a seemingly invincible enemy army. The weapons chosen for the midnight raid are the strangest that ever were: trumpets in their right hands, and in their left hands they each held an old clay jug upside down over a burning torch. As they followed their leaders down from the hills into the valley of the enemy's camp, they blew the trumpets and smashed the earthen vessels. God used the sudden surprise display to totally confuse the enemy and bring His people into a glorious victory.

Perhaps you already see the remarkable parallel.

THE ULTIMATE VICTORY

"We have this treasure in earthen vessels..." Perhaps you've noticed a certain brightness on the countenance of a true Christian. There seems to be a gleam of God's grace that literally brightens the room when they enter. For most of God's people down through time, the only way for the final glory of His presence to be revealed is through the death of this earthly "tabernacle." However, there is coming a day when God Himself will lift the veil and translate His living saints from the vile realm of our earthly estate into the glorious presence of the King of kings and Lord of lords.

OUR ASSIGNMENT FOR THE PRESENT TIME

In a figurative way Jesus told those who follow Him: "You are the light of the world." (Matthew 5:13) Since the glory of God is veiled and unseen by those who are blinded in unbelief, we have been given a life-giving commission to change the situation. (Matthew 5:16)

"Let your light so shine before men, that they may see your good works, and glorify your Father which is in heaven."

Unlocking the Mysteries of CREATION

References For Section Three

1. Jochmans, J.R., Strange Relics From The Depths Of The Earth, Forgotten Ages Research Society, Box 82863, Lincoln, Neb. 1979.
2. Ibid.
3. Ibid.
4. Ibid.
5. Barnes, F.A., "The Case Of The Bones In Stone," Desert Magazine, v. 38, p. 36-39, February 1975.
6. Swanson, Ralph, "A Recently Living Plesiosaur Found?", Creation Research Society Quarterly, v. 15:1, June 1978, p. 8.
7. Mackal, Roy P., "Dinosaurs, Dead or Alive," Ranger Rick magazine, an issue between 1980-1983, p. 12-15.

Selected Reading

Baugh, Carl E., Dinosaur, Promise Publishing, Orange, CA, 1987.
Bowden, M., Ape Men: Fact or Fallacy?, Sovereign Publ., Bromley, Kent, England, 1978.
Gish, Duane, Evolution: The Fossils Say No, Creation-Life Publishers, San Diego, 1978.
Corliss, William R., Ancient Man: A Handbook Of Puzzling Artifacts, The Sourcebook Project, Box 107, Glen Arm, MD. 21057.
Fenton, Carroll L., and Mildred, The Fossil Book, Doubleday and Co., Garden City, N.Y., 1958.
Gish, Duane T., Dinosaurs: Those Terrible Lizards, Creation-Life Publishers, San Diego, 1977.
Morris, John, Tracking Those Incredible Dinosaurs, Creation-Life Publishers, San Diego, 1980.
Taylor, Paul, The Great Dinosaur Mystery And The Bible, CLP, 1986

SESSION 4

UNLOCKING THE MYSTERIES OF
ANCIENT CIVILIZATION

*"FOR INQUIRE, PLEASE,
OF THE FORMER AGE,*

*AND CONSIDER THE THINGS DISCOVERED
BY THEIR FATHERS."*

Job 8:8

Contents of Section Four

NOTHING THEY PLAN TO DO WILL BE IMPOSSIBLE FOR THEM...

The Truth About Man's Primitive Past

WHY ARE ANCIENT CULTURES MYSTERIOUS?

There are many fascinating books and articles being circulated these days with titles that include the word "mystery." **The marvelous findings of ancient civilizations are so surprising, they leave many puzzles apparently unanswered.**

Have you ever stopped to think just why so many things turning up about ancient cultures end up being classed as "mysteries?"

WHAT IS THE TRADITIONALLY ACCEPTED CONCEPT OF MAN'S DEVELOPMENT ON EARTH?

The popular evolutionary scheme regarding man assumes some unproven things to be true.

● Man supposedly began as an ignorant, brutish animal.

● The family of man is assumed to have gradually progressed through the "cave man" stage to become modern man, through a long descendency over millions of years.

● The most ancient beginnings of man's culture are automatically thought to be "primitive" or crude in comparison to later developments.

● Objects produced by very early humans are supposed to be "primitive."

● According to evolutionary notions, anything that is discovered to be both very ancient and technologically or culturally advanced, is automatically labeled as a mystery.

THINK! Is it true that:

ANCIENT = PRIMITIVE ???

SOME FACTS TO KEEP IN MIND

1. As we learned in our discussion on "missing links," there is **no evidence** to even suggest, let alone prove, **that man has developed from any form of beast-like creature.**

The MISSING LINKS of EVOLUTION ARE STILL MISSING!

2. As we learned in section two, **THERE IS NO EVIDENCE TO PROVE THE EARTH IS MILLIONS OR BILLIONS OF YEARS OLD!**

● Such an idea is a fabrication of evolution theory.

● Many strong evidences point to a very young age for our Earth, in terms of a few thousand years, not millions.

THE EARTH IS VERY YOUNG!

EVOLUTION AND MAN'S ACHIEVEMENTS

According to evolutionary thinking, man began at his lowest and gradually evolved upward in technological skills as well as physiological abilities. Of course this supposedly happened over a vast span of millions of years.

THE BIBLICAL VIEW OF ORIGINAL MAN

What does the Bible have to say about ancient man? Psalm 8:5 speaks very clearly that man was made by God just a little lower than the angels. The original word translated here as "angels" is the Hebrew word "ELO-HIM." It is used throught the Old Testament to refer to Almighty God Himself.

According to the Bible, man began at the top and then fell into depravity because of rebellion (sin) against his Creator.

The reality of all verified history, including our most trustworthy history book ever (the Bible), depicts man at his best in the beginning. He has been on a downhill slide ever since.

WHAT DOES ALL VERIFIED HISTORY TELL US ABOUT EARLY MAN?

Keep in mind that our goal is to discover the facts! We want to get information that is verified. We are interested in theories only insofar as they can be backed up by the facts.

The well-known fact of man's existence on Earth is that his history goes back only about 5,000 years. The ancient Hebrew scriptures are not the only record of this. Other early cultures testify to a similar starting point in time.

Unlocking the Mysteries of
CREATION

HOW DOES THE BIBLE RELATE TO ANCIENT CULTURES?

Throughout human history man has repeatedly risen to high peaks of civilization. These great epochs of man's past are acknowledged in the Bible. Abraham's home city of Ur in Mesopotamia was a significant and sophisticated cultural center. The Egyptian dynasty of Moses' time was a magnificent high point in human achievement. Babylon in the days of Nebuchadnezzar and the prophet Daniel was one of the most proud and accomplished societies that ever lived on Earth.

The Bible mentions a number of other ancient cultures whose splendor has only recently been revealed by the archeologist's spade. The Hittites, the Phoenecians, the Syrians and the Persians are a few of the well-known cultures that have come to light in more modern times.

Of course there were many other ancient civilizations on Earth which are not mentioned in the Bible. Keep in mind that the Bible never intended to give a complete history of the world.

WHAT CHARACTERIZED ANCIENT HIGH CULTURES?

Besides the fact that advanced ancient civilizations achieved tremendous **material accomplishments,** there is another quality that is consistently revealed about these peoples. The rulers of these great cultures were involved in what could Biblically be called an **"anti-Christ"** system of **despotism** each time.

When you carefully examine the physical remains of these cultures you will find evidence of widespread **violence** and gross immorality. At the highest peaks of these societies there were **demonic manifestations,** and **human sacrifice** was commonplace.

WHAT HAS BEEN THE END OF EVERY ANCIENT CIVILIZATION?

Virtually every civilization of the ancient past reached a point of destruction and ruin. Despite their great accomplishments, their submission to spiritual wickedness in high places ultimately brought about their downfall.

The Bible clearly records that God has had to bring about catastrophic judgements to finally put an end to the degradation and violence. Other cultures around the world have also been destroyed, sometimes by a mysterious catastrophe.

Even though mankind's depraved cultures have been wiped out, man himself has always managed to rebuild again somewhere. Knowing man's inner capacity and his drive to dominate, it is easy to realize why cultural revival has always taken place down through the centuries.

KNOWLEDGE WILL INCREASE IN THE LAST DAYS

In the Bible we are told by the prophet Daniel (chapter 12 verse 4) that knowledge will increase in the last days. Daniel predicted this nearly 3,000 years ago, at the height of one of the world's greatest cultures. The Babylonian Empire was majestic beyond our comprehension. Early descriptions of the great city leave us in awe at the accomplishments of that high civilization. Mankind has certainly experienced some very low points since then, not the least of which was the period of the dark ages, lasting hundreds of years. But man has indeed managed to again reach great heights of achievement in the last few generations.

WHAT ABOUT CAVE MEN?

As we emphasize the great cultures of the past we must keep in mind that primitive, stone-age people have also existed in remote

parts of the world. Primitive tribes of men have existed in the past just as they do today. They were not the first humans, but **degenerate offshoots of main-line human culture.**

Stone age tribes are living in the jungles of New Guinea and South America today while modern man explores outer space. In similar ways people have managed to scratch out an existence in caves of the past while other humans were advancing both culturally and technologically.

DOES THE BIBLE MENTION CAVE MEN?

The prophet Isaiah observed an interesting thing 2,600 years ago. Notice Isaiah 2:19.

"And men will go into caves of the rocks and into the holes of the ground before the terror of the Lord and before the splendor of His majesty when He arises to make the Earth tremble."

Naturally, in the face of devastating judgement, people who survive will find shelter wherever possible. Caves would be likely habitations for perhaps many years. What most people today fail to realize is that there have been several widespread catastrophes in the ancient past that have literally reduced civilization to rubble. [These catastrophes of the past are studied in detail in volume two of **Unlocking The Mysteries Of Creation.**]

What Were The Capacities Of Original Man?

The Bible boldly declares that God made man in His image—in God's class. Man was intended to be the dominator—a peaceful majestic king over all the creation. To better appreciate the implications of such a position, and the potential of man in his beginning we need to ask a provocative question.

When God created the angels, how much time do you suppose it took before they were fully functional?

It's no problem at all for us to recognize that the angels of God operated on full potential at once, from the beginning of their existence. But what do we think about man? Our concepts of early man have been severely handicapped by our not thinking from the foundation of God's Word.

Did the earliest humans have to grope around idly in ignorance for hundreds or thousands of years before they finally woke up enough to figure out things like fire and the wheel?

JUST HOW SMART WAS ORIGINAL MAN?

We're told that a full grown human only uses about 10% of his brain capacity in his lifetime. In other words, we have ten times more brain potential than we'll ever use.

THINK! If evolution were true, why on Earth is the other 90% of our brain even there?

According to traditional evolutionary theory, body parts are supposed to evolve because of a need for them. This is obviously another one of those *mysteries* of evolution.

AN INTELLECTUAL EXERCISE FOR ORIGINAL MAN

There is one event at the dawn of human history that gives us a significant insight to the mental ability of the first man. The Creator gave Adam the commission to name all the animals. Don't you think they were meaningful names? How would we do with such a job? Judging by the way we name our pets, with words like spot, muffin, and rin- tin-tin, one gets the feeling that modern imagination has suffered some severe setbacks.

Some think that Adam may have spoken in an early form of Hebrew language. Whatever language he used, the names he gave the animals must have been every bit as colorful and significant as the names we use in our own language. The nobility of the lion, the delicacy of the flamingo, the size of the elephant, and the cunning of the fox were no doubt reflected by the names Adam gave. We know they were acceptable names because God accepted them without question. In Genesis 2:19 it says: "And whatever Adam called each living creature, that was its name." God saw no reason to quibble with Adam about his choice of words.

Unlocking the Mysteries of CREATION

Is Technology A Modern Innovation?

When Adam fell into sin, his world fell with him because he was given control of it. But even in his fallen state the family of man accomplished things in ancient times that were far from "primitive."

WHAT DEFINES TECHNOLOGY?

When we think of advanced technology we automatically load it down with all sorts of complicated hardware. The costly things that only a few people know how to tinker with often constitute what we think of as technology. Such things break down easily but we still identify them with the most advanced scientific ways of doing things.

FROM THE VERY BEGINNING, WHY WOULDN'T ANCIENT MAN HAVE ALWAYS BEEN:

- inventive
- technical
- creative
- productive
- ingenious

THE SCIENCE OF AGRICULTURE

In the fourth chapter of Genesis we see that Abel, the first man's son, was a shepherd. His brother, Cain, was a farmer. A descendant of Cain, in verse 20, is described as a forerunner of livestock men.

If you think about it, animal husbandry and agriculture are quite advanced forms of human industry. They require technology in specialized tools and processes. If you think raising cattle or sheep is a primitive pastime, go out and try it for a season. These endeavors are far beyond any supposed "stone age" era of human development.

LONG FORGOTTEN METHODS?

In the *Science Yearbook* of 1980, produced by World Book, an article is titled "Farming The Negev Desert." This arid wilderness in southern Israel is springing to life with lush crops like onions and peaches. Why?

As the lead caption points out, "Israeli scientists have revived long forgotten farming methods to grow crops by storing the sparse desert rain in the once barren soil."

Was this great technological break-through the result of complex, multi-million dollar machinery?

As it turns out, all they did was build little dikes around sloped plant basins to collect the sparse rain. This brilliant achievement involved a minimum of expense and complication. It was an ancient method, but a very productive one.

How would our "modern" technological approach try to accomplish the same result?

To begin with, we would undoubtedly have to construct huge dams hundreds of miles away. We'd build gigantic pumps and pipelines costing millions of shekels to transport the water. Of course we would engineer high powered hydro-electric generators to run the pumping network. Since all of that is useless in dry years the natural modern thing to do would be to build petroleum-burning equipment for a back-up.

The more complex our "modern" technology gets, the more vulnerable we, as a society, become to things like:

- labor strikes

- corporate bankruptcy

- environmental pollution

- monetary inflation

- terrorist sabotage

Our "high tech" society has made us ultra-dependent on a few experts and a few key resources in order to keep the whole system operating.

COULD THE ANCIENTS HAVE HAD BETTER WAYS OF DOING THINGS?

Consider the impact of capitalizing on natural laws to a greater degree than we do. Are there ways to produce larger long-term profits with less complex hardware?

We tend to think that the only kind of technology that can really get the job done is our 20th Century style of industry. But as we explore the clues to ancient peoples we find that they had a different, and possibly wiser approach. For lack of an established definition relating to this area of study consider the implications of a new term:

SIMPLIFIED INGENUITY!

Unlocking the Mysteries of
CREATION

In Genesis 4:17 we find that Cain found his wife, fathered a son, and built a city. Because Cain was the eldest son of Adam and Eve, the first humans on Earth, a question is often asked of Bible- believers: Where did Cain get his wife?

WHERE DID CAIN GET HIS WIFE?

This is a common question. Of course the reason it seems so puzzling is the fact that we all have assumed that Cain and Able grew up together as the only children of the first parents. It would appear on first glance that there must have been some other family of humans somewhere in the vicinity to provide Cain with a wife. Have you ever thought of the alternative to the purely Biblical position?

WHAT'S THE WHOLE STORY?

Take a look at Genesis 5:4 and you discover that Adam fathered other sons and daughters besides the three sons that are mentioned in the Bible (Cain, Abel, and Seth).

It is obvious that children of the first family of humans had to intermarry. Cain would have taken one of his sisters to be his wife.

Some may say to you: "Doesn't God forbid marriage among brothers and sisters?"

When did the taboo against sibling marriages originate? Who was the girl that Abram took for his wife? She was Sarai, his half sister. It wasn't until the time of Moses, about 500 years later that God laid down the law not to marry your brother or sister.

WHERE DID THE EVOLUTIONARY CAIN GET HIS WIFE?

One of the major unanswered mysteries of evolution is the problem of having both a female and a male of the species available at the exact same time and place in order to reproduce a second generation human. Another one of those famous "chances" is invoked here, insisting that there just happened to be a biologically compatible female to accommodate her mate and produce offspring and secure Homo sapiens from becoming extinct right at the start.

But the consideration here is the genetic problem: intermarriage within a family can breed congenital malformations and health problems. We need to keep in mind why this is so. With an accumulation of genetic mutations over many generations the likelihood of a problem because of inbreeding becomes increasingly dangerous. But the closer you get back to the original perfectly created man, the cleaner the genetic "pool" would be.

CITIES BEFORE THE FLOOD?

So Cain built a city. Notice it doesn't say he built a mud hut! This first son of the first human couple built the first city on Earth. That does not mean that his city was something on the order of a city like New York. But whatever its size, this early city would have some basic characteristics of cities in any period of history.

Architectural design would surely have been evident in the houses and shops along the streets and lanes of Cain's city of Enoch.

WHAT DO PEOPLE SAY ABOUT THE OLDEST CITIES ON EARTH?

It's interesting to see the historical comments and magazine article titles highlighting these most ancient cities. It is not uncommon to find titles like these (from various issues of Readers Digest):

- Palenque: Mexico's Mysterious Lost City

- In The Beginning: The Mystery Of Ancient Egypt

- Mysteries Of The Maya

- Unsolved Mysteries Of The Great Pyramid

Social structure would no doubt be in the families and among their relationships with others in the community.

Political organization, a necessary part of every orderly city, must have been found in the leadership as men asserted their ingrained tendency to seek recognition and dominate others.

Some may challenge the idea of cities existing so early in man's history. Actually, as we shall cover shortly, the very earliest relics of human history indicate the existence of cities from the beginning.

THINK! Why do such references to very advanced ancient cities get called "mysteries?"

The area of the world called the Middle East includes what has frequently been labeled the "cradle of civilization." Early empires in the lands of countries like Egypt, Iraq, Syria and parts of the Mediterranean Sea built cities thousands of years ago. The ruins of these places testify to magnificent accomplishments despite their antiquity.

MUSIC

"Jubal . . . was the father of all those who play the harp and flute."

Genesis 4:21

This Sumerian harp is one of the oldest musical instruments in the world. It dates back 3,500 years and is a magnificent piece of workmanship.

Early mankind, before the flood, knew the technical skills involved in the production and use of musical instruments. That may come as a surprise to those who have been taught that music evolved at a much later time.

TRADITIONAL IDEAS

What is the usual picture painted for us about the discovery of something like music. It might go something like this:

Grampa Ramapithecus was out with the family picking berries one day. As he reached for a juicy looking berry dangling overhead, he chanced to stumble over a vine of some sort that was stretched rather tightly between two trees. He just happened to notice the curious twanging sound that he hadn't heard before. He turned back to pluck the vine again. Twa-a-a-ang . . . it did it again. He grunted excitedly to his fellow berry-pickers nearby who had gotten side-tracked picking termites out of the ground and lice out of each other's hair. "Hey, look what I found!" (or grunts to that affect).

Though this is a funny tale, it is probably not too far off the usual explanation for the discovery of things like this.

MUSIC IN ANCIENT CHINA

A Chinese zither, similar to our modern autoharp, was found recently in a tomb dating back over 21 centuries. Also in the tomb was a mouth organ with 22 carefully designed bamboo pipes, a mouthpiece and fingering stops to operate it. The details were finely crafted. In no way could these things be called primitive.[1]

According to the Chinese people there was an emperor living 4,600 years ago who wanted to standardize musical sounds. He sent his servant into the mountains in search of a special bamboo pipe that would make the sound of a C note when cut to an exact specified length. From that pipe all the other notes of the musical scale were then mathematically derived.[2]

METALLURGY

"Tubal-Cain, an instructor of every craftsman in bronze and iron." Genesis 4:22

The iron age is ordinarily thought to have begun a little over a thousand years before Christ. The so-called bronze age supposedly began another 2,000 years before that. But this reference claims that both bronze and iron were manufactured significantly before that, even before the flood.

From the Danube River basin in Europe have come sophisticated copper tools dating back at least 4,000 years

The National Geographic Magazine featured a surprising article in the November 1977 issue titled: "Ancient Europe Is Older Than We Thought." The newest findings are changing ideas about the "barbarians." Quoting from the article: "Scholars once thought that metallurgy spread here from the Near East, but (there is) . . . proof (now) . . . that a metal industry was fully developed in Europe when it was just starting in the Aegean (i.e. Greece)."

ANCIENT METAL WORKS ALL OVER THE WORLD

It has been reported that at the foot of Mount Ararat in Turkey Russians have found hundreds of ancient ruined smelting furnaces. The evidence verifies that early settlers in the area made a variety of alloys of bronze, including the elements: copper, tin, zinc, and arsenic. [3]

In ancient Egypt discoveries of elaborate goldsmith work date back more than 4,000 years ago. Some metal artifacts from ancient Egypt have been found to be electroplated! That is, gold has been applied over base metals through the process of electrolysis. [4]

From a 22-century-old emperor's tomb in China comes this report in National Geographic for April 1978. "Metal swords, (buried for 2,200 years, have been) treated with a preservative that prevented corrosion for (all that time)." The ancient weapons were alloyed of tin, copper, and 13 other elements, including magnesium, nickel, and cobalt.

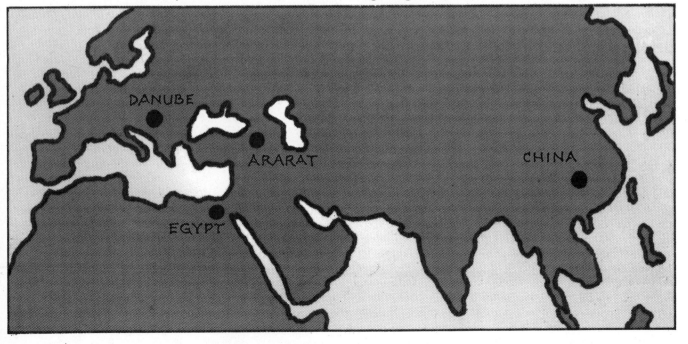

Why Search Out The Former Age?

"For inquire, I pray thee, of the former age, and prepare thyself to the search of their fathers."
Job 8:8 (KJV)

The book of Job is the oldest book of the Bible. It dates back to the time of the patriarchs, 2,000 years before Christ. Job may have lived even before Abraham.

In this brilliant piece of ancient prose, the writer speaks of an even more ancient era with the implication that people before Job's time had far greater knowledge than they did.

"For we are but of yesterday, and know nothing, because our days upon Earth are a shadow."
Job 8:9 (KJV)

HOW LONG DID THE EARLIEST HUMANS LIVE?

When Job contrasts the longevity of his contemporaries with the people who lived in the "former age" he says, "our days . . . are a shadow." what does he mean?

According to the historical record, Job lived just a few hundred years after the world-wide destruction of the Great Flood. Could the "former age" he writes about be referring to the era before that great dividing point of human history?

The flood radically altered Earth's environ-

ment. One of the most significant affects of the change was the reduction in human lifespan. A number of changed environmental factors would have caused this, especially the removal of the canopy that shielded Earth's atmosphere.

A Sampling Of The Length Of Lives Of Men Who Are Mentioned In The Bible

BEFORE THE FLOOD	
Adam	lived 930 years
Seth	lived 912 years
Enosh	lived 905 years
Kenan	lived 910 years
Mahalalel	lived 895 years
Jared	lived 962 years
Enoch	lived 365 years*
Methusaleh	lived 969 years
Lamech	lived 777 years
Noah	lived 950 years**

Enoch (*) did not die because he walked so closely to God that God finally just took him. See Genesis 5:24 and Hebrews 11:5.

Noah (**) lived 600 of his 950 years before the flood.

WHAT COULD PEOPLE DO IN 900 YEARS?

Think of some of the things that you would be able to do if you lived 900 years or so. That is equivalent to 13 of our 70-year life-times.

With that kind of time on your hands it wouldn't be hard at all to take off a hundred years and perfect your piano playing skills.

If you are more prone to developing a

business, can you imagine what an empire you could put together if you had several hundred years to do it? You could spend most of your life enjoying the profits of your work.

Let's say you're the kind of person who likes to travel and explore the world. Think of the places you could discover and enjoy for really extended vacations.

Now keep in mind that these people were every bit as human as we are. And they lived with millions of other humans with just as wide a cross section of interest and talents as you would find in our society today.

In the last few thousand years of man's history his brief longevity has been his most severe handicap. Even at that look at what men have been able to do with their God-given abilities and aptitudes. Alexander the Great conquered the entire Mediterranean world by the age of 30. Look at the hundreds of inventions produced in the single brief life of Thomas Edison. Consider the stacks of musical masterpieces created by the minds of men like Mozart (who died at age 35), Bach (died at 65), and Beethoven (died at 57).

IS IT ANY WONDER WHY JOB URGES HIS READERS (US) TO SEARCH OUT THE KNOWLEDGE OF THE FORMER AGE?

AFTER THE FLOOD

Shem	*lived 600 years**
Arpachs.	*lived 438 years*
Shelah.	*lived 433 years*
Eber	*lived 464 years*
Peleg	*lived 239 years***
Reu	*lived 239 years*
Serug	*lived 230 years*
Nahor.	*lived 148 years*
Terah	*lived 205 years*
Abraham	*lived 175 years*

Shem (*) was only 100 years old when the flood came.

Peleg (**) was born at the time of another world-wide catastrophe (the division of the continents) which dropped his life span, and that of his descendant to about half the age of his post-flood ancestors.

THE COMPOUND VALUE OF SHARING LIFE WITH GENERATIONS OF OTHERS

On the following page you will notice how the lives of pre-flood men overlapped for hundreds of years.

Noah lived 950 years. Six hundred of those years were spent with his grandfather. Do you think they just sat around and played dominoes all that time?

But the story gets even more intriguing when you realize that Noah's grandfather, Methusaleh, likely knew Adam for over 200 years! Think of it; sharing the insights and discoveries of the first man—the one who walked with the Creator in the Garden of Eden and talked with God face to face.

PEOPLE WHO LIVE LONG KNOW MUCH!

"With the ancient is wisdom; and in length of days understanding."

Job 12:12 (KJV)

With all the natural resources available to them why wouldn't man before the flood have exercised his natural human ingenuity every bit as much as he has in the last two thousand years?

Developments in commerce, industry, the arts, politics, science, and medicine would have all been possible before the flood.

Yet how often have you seen traditional Sunday school art depicting Noah using "stone age" tools to build the largest wooden ship in history. Do you see how our concepts of ancient man have been groomed by evolutionary brainwashing?

Imagine The Accumulation of Ancient Wisdom Over The Generations

YEARS AFTER CREATION

| 0 | 100 | 200 | 300 | 400 | 500 | 600 | 700 | 800 | 900 | 1000 | 1100 |

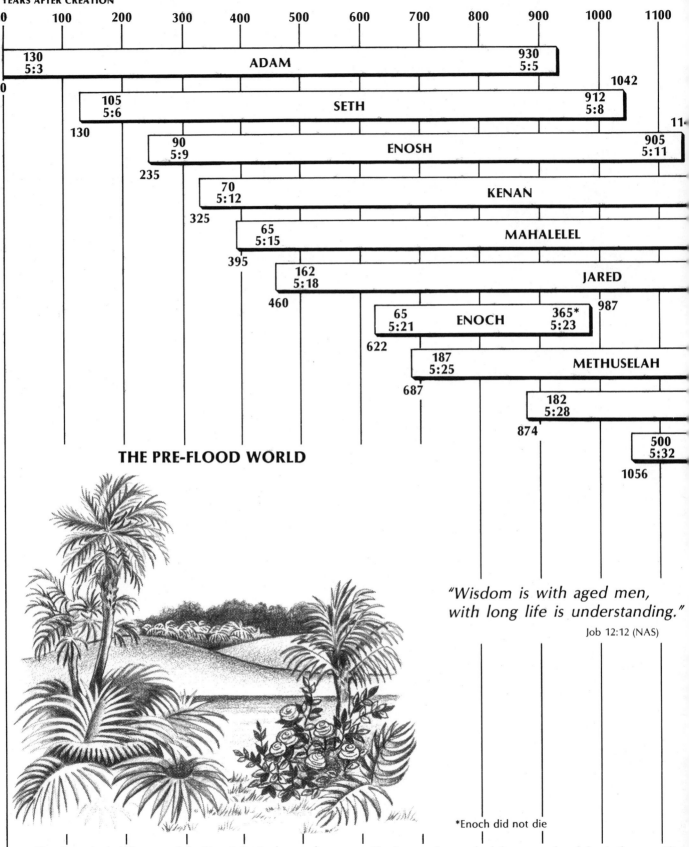

ADAM — 130 / 5:3 ... 930 / 5:5

0

SETH — 105 / 5:6 ... 912 / 5:8 — 1042

130

ENOSH — 90 / 5:9 ... 905 / 5:11

235

KENAN — 70 / 5:12

325

MAHALELEL — 65 / 5:15

395

JARED — 162 / 5:18 ... 987

460

ENOCH — 65 / 5:21 ... 365* / 5:23

622

METHUSELAH — 187 / 5:25

687

182 / 5:28

874

500 / 5:32

1056

THE PRE-FLOOD WORLD

"Wisdom is with aged men, with long life is understanding."

Job 12:12 (NAS)

*Enoch did not die

The chronological life spans of the Genesis Patriarchs reveal many surprising facts.

HOW TO USE THE CHART

• The vertical lines mark off centuries beginning with Adam's creation in the year zero.

• The bar for each individual represents his total length of life compared to the others in the chart.

• The first number, at the left corner of each bar is the year After Creation (A.C.) of the man's birth. (Enosh: born 235 A.C.)

• Inside each bar the first numbers from the left show the age at which he fathered the son beneath him, and the Bible verse where that information is found.

• At the end of each bar the top number is the year After Creation of death. The number beneath is the age at death. Then the Bible verse in Genesis is listed.

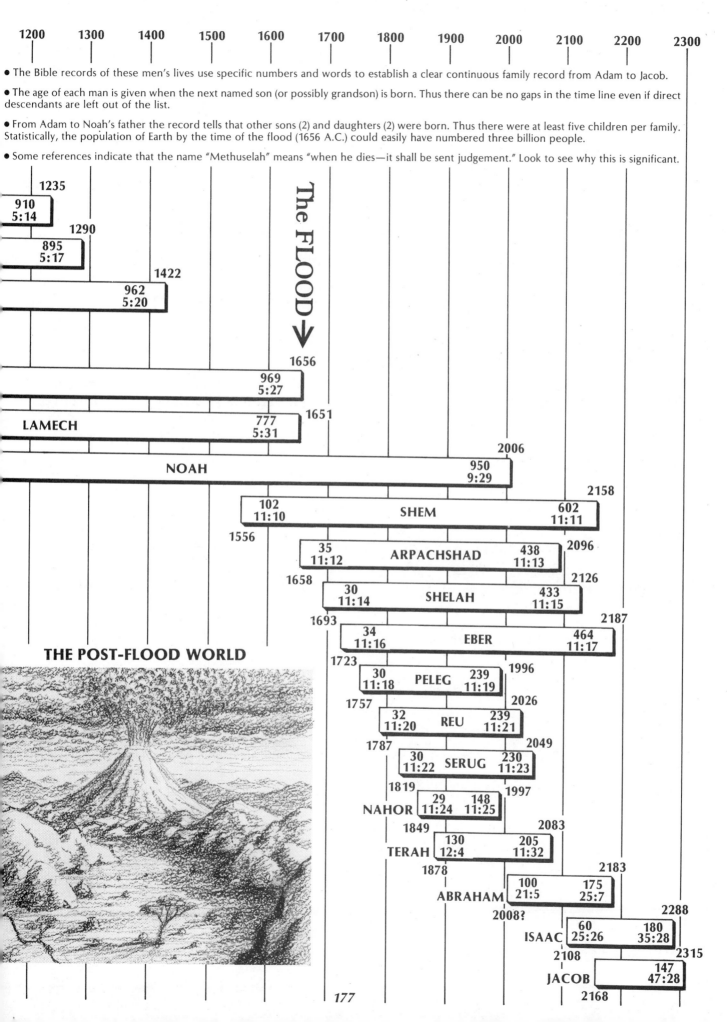

- The Bible records of these men's lives use specific numbers and words to establish a clear continuous family record from Adam to Jacob.

- The age of each man is given when the next named son (or possibly grandson) is born. Thus there can be no gaps in the time line even if direct descendants are left out of the list.

- From Adam to Noah's father the record tells that other sons (2) and daughters (2) were born. Thus there were at least five children per family. Statistically, the population of Earth by the time of the flood (1656 A.C.) could easily have numbered three billion people.

- Some references indicate that the name "Methuselah" means "when he dies—it shall be sent judgement." Look to see why this is significant.

The FLOOD →

THE POST-FLOOD WORLD

JOB'S THEOLOGY

Even Job, in the early centuries after the flood, had amazing insights. His book verifies he had considerable knowledge of agriculture, building trades, and the laws of nature. But he had amazing insights into spiritual truth too.

JOB HAD SOMETHING IMPORTANT TO SAY

"Oh, that my words were written! Oh, that they were inscribed in a book!"

Job 19:23

Skeptics used to doubt that even Moses could write intelligently. Here we see a man who lived some 500 years before Moses, and he's contemplating printing a book! But his message is so important he also wants to emblazon it on granite.

"That they (my words) were engraved on a rock with an iron pen and lead, forever!"

Job 19:24

In Jeremiah 17:1 it says that the sins of Judah are inscribed with an iron pen with the point of a diamond. Is it possible that 20th Century industry is not the first to have diamond-tipped iron cutting tools? Obviously he wants to make a very indelible record and share this bit of knowledge with others for a long time to come.

WHAT IS JOB'S IMPORTANT MESSAGE?

" . . . I know that my Redeemer lives, and at the last He will take His stand on the Earth."

Job 19:25 (NAS)

Hold it! How did this ancient man understand such deep theology even before Moses and the Bible?

One of the feeble arguments among Bible critics maintains that man invented the idea of God and the supernatural only after he evolved enough to produce such sophisticated thought.

The very idea of "redemption" is a complex and advanced element of systematic theology. The idea of a coming Redeemer was familiar to Job because he had a continuous link of ancestors back to the man who knew God in the Garden. Job, a man who lived 4,000 years ago, had insights that stun modern anti-Biblical philosophers.

In verse 26 Job goes on to say that he knew he would see God from his own resurrected body! Such a statement shows incredible insight, far beyond what would be expected of a primitive tribesman.

WHERE DID JOB GET THESE IDEAS?

Look at the chart on the preceeding page and it becomes quite apparent that families could easily pass down and reinforce important information. Noah's family carried a tremendous link of traditional knowledge from the old preflood world to the new age in which Job lived.

Noah's grandfather, Methuselah, no doubt told him about his father Enoch who left this world nearly 70 years before Noah was born.

In Jude 14 and 15 we read that this contemporary of Adam (Enoch) predicted the "coming of the Lord with ten thousands of his saints to execute judgement on wicked mankind in the last days."

Yes, ancient man had a keen insight on spiritual realities.

WHAT DID ORIGINAL MAN HAVE?

By now you have no doubt begun to add some new dimensions to your understanding of the original man and his vast family. The Bible does put things in new perspective doesn't it? As it says in Psalm 119:130, "The entrance of Thy words give light . . . "

It's an amazing thing to realize that the first man actually shared intimate fellowship with the Creator. **Original man was so totally different from the secular impressions we are subtly brainwashed into believing.**

The following are just a few of the superior qualities enjoyed by original man:

- complete harmony with nature
- perfect communication with God
- peaceful dominion over the animals
- intimate awareness of natural laws
- intimate response to spiritual laws

God's masterpiece of creation, man, was made to be the master of the entire Earth. He didn't have to be some phony kind of "superman" in physical strength in order to do it either. With his God- like spirit, man had the power to accomplish anything to which he applied his ingenious mind and corporate energy.

Even after the fall in sin, God said of mankind:

"Nothing they plan to do will be impossible for them."

Genesis 11:6 (NIV)

Keep in mind that this statement was made by God himself, even years after the flood, at the time of the tower of Babel incident.

That may seem like quite a sweeping statement for God to make about man. But keep in mind how God made man in the beginning. He was in God's class, made "just a little lower than Elohim (God) Himself."

Now let's look for the evidence to support this Biblically realistic way of looking at human accomplishments.

What's The Solution To The Puzzle Of Advanced Ancient Technology?

As modern man has probed around his world, he has made many fascinating discoveries. Ancient relics show evidence of high degrees of culture and technology. In our time these puzzling discoveries have shown ancient people to be advanced far beyond what was traditionally expected.

Classic examples include:

● The accurately placed megaliths (large cut stones) at Stonehenge, England.

● The Great Pyramid at Giza in Egypt.

● Tremendous ancient cities like Babylon and Teotihuacan in Mexico.

● Ancient knowledge of mathematics and astronomy, as at the Mayan center of Chichen Itza.

● Even curious hints of skilled tools and machinery, electricity, and air power.

ONE POPULAR EXPLANATION

In our generation the discoveries of ancient technological wonders have been exploited by dreamers like Erik VanDaniken in his familiar book, *Chariots Of The Gods.*

Their idea is that since Earthlings must surely have been too primitive to do all this on their own, the obvious explanation is that superior alien beings brought the technology here from distant worlds in outer space. "The absurdity of this premise becomes very clear in light of scripture and known astronomical observations."

WHAT IS THE BIBLE'S PERSPECTIVE ON ANCIENT TECHNICAL ACHIEVEMENTS?

To gather insight, let's explore the wisdom of the wisest man (outside of the Messiah) who ever lived. He lived in Israel a thousand years before Christ. He was the king of Israel for forty years during the highest golden age of the nation's history. In the book of First Kings 3:12 we find that Solomon was supernaturally gifted by God with a wise and understanding heart. There was none like him before or after that time. Besides being a ruler, he was also a naturalist and writer who studied all manner of things.

His essay is called Ecclesiastes in the Bible. Let's examine a part of Solomon's introduction and discover one of his keys of understanding the real world.

"That which has been is what will be; that which is done is what will be done, and there is nothing new under the sun.

"Is there anything of which it may be said, 'See, this is new'? It has already been in ancient times before us.

"There is no remembrance of former things, nor will there be any remembrance of things that are to come by those who will come after."

Ecclesiastes 1:9-11

Just think what Solomon is saying to us from a vantage point of 3,000 years ago. Remember also that he lived 1,000 years after Abraham, and 3,000 years after God created Adam. Solomon is giving us a perspective on human achievement from the very center point of all human history.

There's nothing new under the sun. It's all been done before. He's talking about *things* isn't he? Things that man has invented or discovered and adapted for his use. Solomon says that all the things we might call "new" have already existed in "ancient" times.

Now apply that from our perspective. Solomon seems like an "ancient" to us, but here he's calling other more ancient people the ones who have done it all before. Do you think he's referring here to the citizens of that "former age" that Job mentioned?

The wise king observes that people are very forgetful. Maybe that's because we don't live long enough anymore. It is true though isn't it? Just think how many things have been forgotten in "modern" times. There used to be many "coopers" in any town, making wooden barrels for shipping just about everything. Handmade barrels were the cardboard boxes of the colonial days. But now do you suppose you could find a cooper? Barrel making is almost a lost art. Isn't that exactly what Solomon predicted? "no remembrance of things to come by those who will come after."

Just when we really think we've "arrived" with some "state-of-the-art" discovery, someone comes along and, in a few years, makes our great invention obsolete. Are we simply repeating the discoveries and inventions of another age, a forgotten society which could even have surpassed our own?

Remember what God made man to be. Even in fallen separation from the Creator, God said "nothing they plan to do will be impossible for them." (Genesis 11:6)

WHAT IF THERE WAS A GLOBAL FLOOD?

If the great flood did, indeed, obliterate everything on the Earth about 5,000 years ago, it stands to reason that the oldest records of civilization would only go back that far!

It is interesting that in fact, advanced cultures on Earth seem to appear suddenly at about that time. Even more revealing is the fact that these most ancient advanced cultures begin in an advanced state, without a trace of "primitive" generations leading up to them.

THE AMAZING EARLIEST RELICS

The Great Pyramid Of Cheops

Near Cairo, Egypt this magnificent structure is one of the oldest buildings on Earth. In terms of its sheer mass, it may be the largest single building in the world. Curiously, many have acknowledged that this lone-survivor of the seven man-made wonders of the world, is also the most perfect building on the planet!

The intricacy and stability of this monument of human engineering are absolutely phenomenal. It's hard to imagine the detailed planning necessary to produce such a massive structure, and not have even the foundation sag in over 4,000 years!

The earliest legends of the surrounding people insist that the forefathers of the builders were even more knowledgeable. There is no evidence here of a primitive past. These people had a sophisticated religious and social order.

The May 1975 issue of *Readers Digest* featured a thought-provoking article entitled, "When Did Civilization Begin?" One statement in it is particularly worth noting:

"the new findings have made a shambles of the traditional theory of prehistory."

THINK! Could it be that man did not really start at the bottom of the evolutionary scale?

The Minoan Civilization

On the Mediterranean island of Crete a very modern civilization was completely devastated by an immense volcanic eruption 3,500 years ago!

The reconstructed Palace of Minos verifies the excellent architectural ability of the Minoan people.

The ancient Minoans were highly advanced in many things including:

- government
- the arts
- social structure
- language
- medicine
- mathematics
- astronomy
- sea travel

The sophistication of their city looks modern with bright columns, polished marble floors, and intricate ceiling decorations. Their buildings were every bit as fine as many of our highly crafted government buildings today. Their delicate pottery rivals the finest of any culture in any historical time.

DISCOVERIES THAT STARTLED THE WORLD

The *National Geographic* magazine of February 1978 includes the amazing reports of Crete's early archeologists. Heinrich Schlieman in 1873 and Sir Arthur Evans in 1900 were the men in modern time to reveal the surprises of Cretan culture from as early as 1650 B.C.

The impressions of these modern explorers are interesting in that they reflect an obvious preconceived stereotype about ancient man. Notice what they said, as quoted in *National Geographic*.

"A remarkably sophisticated society for so distant a time."

THINK! Why would sophistication seem "remarkable" in a context of antiquity? Would such a comment be made by a person with the insight and knowledge of men like Solomon and Job?

When Was The Golden Age?

When you think about the great accomplishments of human civilization down through time, consider the question: When was the Golden Age for all the various major cultural centers on Earth?

GREECE
When was the golden age of Greece?
Athens 500 B.C.

MESOPOTAMIA
When was the golden age of Mesopotamia?
Babylon 500 B.C.

MEXICO
When was the golden age of Mexico?
Mayan centers 500 B.C. - 500 A.D.

ITALY
When was the golden age of Italy?
Rome 200 B.C. - 200 A.D.

CHINA
When was the golden age of China?
Dynasties of Emperors Before Christ

All the Golden Ages are past! Most of the people living in these places today are experiencing anything but splendor. The heights of cultural achievement are long gone. Is civilization at its creative best today? Or are forgotten civilizations of antiquity proving they were superior?

EGYPT
When was the golden age of Egypt?
Dynasties of Pharoahs 1500 B.C.

DID ANCIENT MAN TRAVERSE THE OCEANS?

For years the modern western mind-set held the view that Columbus discovered America in 1492. Later it was conceded that Nordic Vikings from Scandinavia reached North American shores about 1000 A.D. The implication has been that man's ability for global travel has only been developed in relatively more modern times.

But if ancient man was as capable as we are projecting here, then why shouldn't we expect to find evidence that they traveled all over the globe?

The March 1973 issue of *Readers Digest* magazine included an article entitled: "Who Really Discovered The New World?" The article surprised readers with the following statement:

"startling evidence reveals that the American continent drew many early visitors, including some more than 2,000 years before Columbus!"

A SURPRISING ANCIENT MAP

One of the most well publicized and puzzling artifacts relating to this subject is the famous Piri Reis Map. Historically, this antique map dates back to the year 1513. However it was copied from other more ancient maps dating back even earlier, possibly as far as the period before Christ.₅

So, what is so surprising about that?

The curious thing about the Piri Reis Map is that it includes the portion of our planet we call Antarctica, the region around the South Pole. Even more unusual is the fact that it outlines the actual continental coastline of that land. Of course the tremendous ice pack covering Antarctica extends much farther than the land coastline.

We realize that it was not until the year 1952, by modern technical means, that the land coastline of Antarctica was charted. How could it have been charted by ancient mariners prior to the year 1513? The problem of course is in our limiting ideas about the technical abilities of ancient man. The point is simply this: Ancient man did accomplish many outstanding things because he was indeed able to do so. The means by which he did such things are a mystery because such knowledge was forgotten. And that is exactly as Solomon predicted it would be!

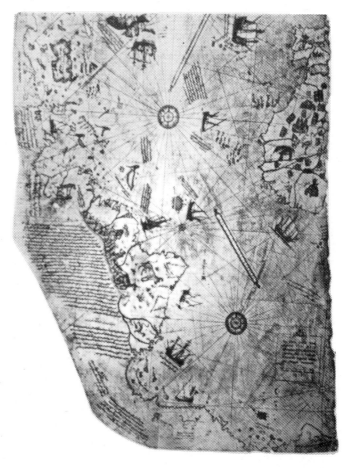

The Piri Reis map

AMERICA BEFORE CHRIST - WHO VISITED?

An outstanding book to shed light on the subject of early global travel is *America B.C.,* by Harvard University professor Barry Fell. Doctor Fell graphically documents archeological finds that show us there were many global travelers before the time of Christ.

He reveals that North and South America received ancient visitors from Europe and North Africa. Discoveries in Canada, the U.S.A., Central America, and South America show that many trans-oceanic contacts were made with people called Celts, Iberians, Egyptians, and Libyans.

One example of the kind of thing that has been discovered is a stone inscription found in West Virginia in the year 1838.[6] The writing on the stone could not be deciphered at the time, but now it has been determined to be a Phoenician language that was used in Spain before Christ. How did it manage to get to West Virginia? Apparently, it was left there by a visitor who arrived long before the American colonists.

Ancient stone monuments have also been found in a variety of places in the Eastern U.S.A. These giant stones were set up by foreign men who traveled across the Atlantic Ocean possibly as much as 3,000 years ago.

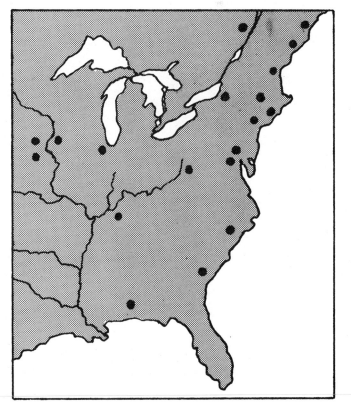

As shown on this map, there are many sites in Eastern North America where Old World artifacts have been found.

Ancient Sea Travel

Did ancient civilizations travel to diverse parts of the globe? The traditional understanding of ancient people says "no." Evolutionary thinking about man's technological development insists that such advances are only the domain of relatively recent and more "modern" man. But what does the evidence show us?

THE ANCIENT POLYNESIANS

The ancient Polynesian people are typically classified as a rather primitive society. Yet they are famous for their vast travels all over the mighty Pacific Ocean.

By the time of Christ these aggressive explorers had colonized every habitable island in the Pacific Ocean. They had methodically covered over 15 million square miles of open sea.[7] These mariners knew how to navigate by the sightings of the stars. They also had techniques to understand the currents of the ocean and the barely perceptible ocean swells. Both of these phenomena have been discovered by western civilization only in what we call "modern time."

MYSTERIOUS EASTER ISLAND

One of the places inhabited by the Polynesians is Easter Island. It has become well-known for its megalithic heads carved out by a stone-age society.

These great monuments have been called mysteries by modern scientists. Some of the massive heads, smoothly sculptured out of iron-hard volcanic rock, weigh over 180,000 pounds (90 tons). About 2,300 miles off the coast of Chile, this remote barren island is populated by at least 1,000 of these strange giants. Most of them are 12 to 15 feet in height. One is 32 feet tall.[8] How did they get here?

Transported several miles from the quarry to their platforms, these stone-age masterpieces are a puzzle to modern researchers. How were they carved and moved? Their immensity is a challenge even to modern engineers. What was the purpose of these strange figures, gazing soberly across windswept, treeless plains? The society that made them is gone forever. But what they left behind makes us wonder if they knew much more than we usually credit to them.

THE WORLD OF MEXICO BEFORE COLUMBUS

There is more evidence of early world travel that comes from the ancient Mayan centers of Mexico and Central America.

Ceramic heads testify clearly that people of different racial groups had contact with the ancient inhabitants of Mexico.

Numerous ceramic and stone artifacts show clear characteristics of African and Oriental people. Even Phoenician features show up on an ancient ceramic cup from Guatemala.

The amazing absurdity is that such evidence means nothing to those "experts" who deny the possibility of ancient world-wide travel. Many of them insist that the only way humans ever could have arrived here is by slow migration across the Bering land bridge from Asia to Alaska. Down through the years these post-ice-age people supposedly inched their way southward, finally to settle in Mexico and begin the development of the highly advanced Mayan culture.

DOES THE BIBLE OFFER ANY INSIGHT ABOUT GLOBAL SEA TRAVEL?

In the Old Testament book of First Kings (10:22) we read about King Solomon's shipping enterprise. This great king joined forces with neighboring King Hiram of Phoenicia. Together they sent merchant ships on voyages lasting three years. They brought back all sorts of exotic imports from distant lands.

The Phoenecian neighbors of King Solomon have been called the Sea-Lords of Antiquity. They were masters of the ocean. We know they traveled all over the Mediterranean world and even around the coast of Africa. But why do people think they couldn't make it across the Atlantic Ocean?

Looking at a map of the world you will notice that the distance from Africa to South America at their closest points is actually less than the length of the Mediterranean Sea by at least twenty five percent. If ancient sailors could expertly navigate the Mediterranean, it would be a cinch to get across the Atlantic.

Comparing the Atlantic to the Pacific Ocean one easily sees that the Atlantic is a mere lake by contrast. If the Polynesians mastered the Pacific over 2,000 years ago, why couldn't the mighty Phoenicians manage to get across the tiny Atlantic? Obviously, the only barrier is not the ocean, but rather the preconceived bias that ancient people were too backward to ever do something so advanced.

FROM PAINTING BY LLOYD K. TOWNSEND

Why Are Ancient Pyramids A World-wide Phenomena?

ANCIENT BABYLON

The famous Babylon we read about in the Bible is in present-day Iraq. The land has a history dating back all the way to the great flood.

The Tower of Babel was very likely a pyramid or ziggurat structure, possibly the first one to be built after the flood.

The ruins of several ancient ziggurats still exist in this arid part of the world. The one at the ancient site of Ur (Abraham's birthplace) dates back over 4,000 years to 2100 B.C. They are a silent witness to the magnificence of that ancient civilization.

Also from ancient Mesopotamia a strange and puzzling object is sure to arouse our curiosity. It is found in the Baghdad museum and has been publicized in several books.

A ceramic jar was found with a copper cylinder suspended from its mouth into the center of the jar's cavity. Inserted into the length of the cylinder there was an iron core. When mere grape juice was put in the jar along with the cylinder and core, it was found to produce an electric current of one and a half volts![9] Here we thought Benjamin Franklin discovered electricity while flying his kite in a rainstorm. Yet it is now known that ancient people used electricity because electroplated relics have been found dating back to 2500 B.C.[10]

ANCIENT EGYPT

Pyramids are usually associated with ancient Egypt. The great pyramid of Cheops near Cairo was already a 500-year-old antique when Abraham and Sarah arrived there from Ur of the Chaldees. The oldest pyramid at Saqqara dates back another 200 years. Historically, this would have been very close to the time of the great flood. Some believe that the great pyramid is the oldest, and that it dates back to before the flood. The Egyptian pyramids have been shrouded in mystery. There are many differing opinions about their possible origin, purpose, and method of construction.

It is often a surprise to people when they learn that Egypt has only a few pyramids compared to the number of similar structures in Central America.

ANCIENT MEXICO

The magnificent Pyramid of the Sun near Mexico City is an awesome structure 216 feet high. As tall as a twenty-one story office building, this man-made mountain of rock and brick is almost half the height of the Great Pyramid in Egypt. It is curious to note that the base dimensions of these widely separate structures are almost the same. (Egypt's is 750 feet square; Mexico's is 720 feet by 760 feet.) Does that seem too close to be coincidental? Could there be a similar source of design, even though the two structures are a third of the way around the world from each other?

Mexico and Central America are loaded with ancient pyramids. Ancient Mayan settlements number over 1,100 centers found so far. Each center usually has a dozen or more

similar in design to the pyramid-tombs of the Egyptian pharoahs.

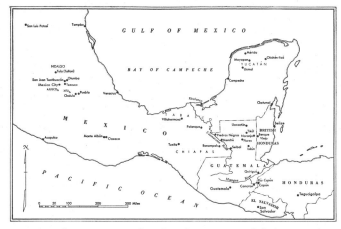

pyramids. Some think there could be over 100,000 pyramids overgrown by the jungles of Mexico.

The ancient Mayan center of Tikal is one of those splendid ruined cities. It is overtaken by a dense jungle now and only part of it has been exposed by archeologists. But is is known to have once covered an area of 50 square miles!

Palenque is another amazing city which displays evidence of a very advanced culture in the middle of the steaming jungle. The tomb of Pacal, an ancient ruler, is strikingly

El Tajin is an interesting pyramid because it's four sides are covered with little niches. There are 365 of them altogether; one for every day of the year.

Teotihuacan, the majestic Mayan city where the Pyramid of the Sun is found, covered eight square miles. Legends say it was built by giant, white demi-gods. It is an intricately layed-out city with many massive stone buildings. As many as 200,000 people are thought to have lived here. This civilization flourished for a thousand years before it fell apart in the eighth century A.D. before the Aztecs came into dominance in that part of the world.

MAYAN ASTRONOMY

The ancient Mayans of Central America had a keen awareness of the cosmos. They built observatories like the one found at Chichen Itza in the Yucatan peninsula. The entire culture of the Mayan people seems to have been focused on the watching of the stars and planets.

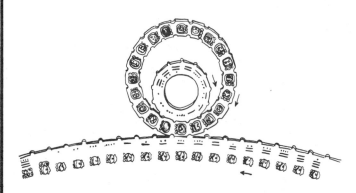

These ancient astronomers devised an accurate cosmic clock to predict the solstices and equinoxes (first days of the four seasons).

Their precise mathematics and astronomical calculations enabled them to figure the length of a solar year to within one ten-thousandth of a day. The Mayan year was 365.2420 days long. It has been only recently that modern astronomers have been able to calculate the solar year to be 365.2422 days.[11]

The orbit of the planet Venus was monitored closely by the ancient inhabitants of Mexico. The famous Aztec calendar was specially integrated with the cycles of Venus. There was a good reason for this. According to these ancients the planet Venus was responsible for periodic catastrophes that could desolate their society every 52 years. These astronomical cycles were woven into their religious and cultural patterns.

Our modern intellectual community tends to dismiss ancient religious practices as mere ritual with no basis in fact. This is because humanists deny the reality of the spiritual world. Consequently, anything sounding like supernatural judgement coming out of the heavens is ridiculed as unrealistic mythology. In the upcoming second volume of this seminar series we will explore much more about ancient heavenly catastrophes which destroyed entire civilizations.

THE GARGANTUAN STONES OF TIAHUANACO

High in the Andes Mountains of Bolivia, are the stark and mysterious ruins of Tiahuanaco. This obscure desolation sits at an altitude of 13,000 feet. It seems impossible that an inhabited city like this could be positioned where the air is so thin. Yet this mysterious city may help us to better appreciate the abilities of the ancients.

Almost nothing is known about the history of this place. Who lived there? When did it flourish? How far did its empire extend?

In the book, *The World's Last Mysteries,* by Readers Digest, the lead headline reads: "When the winds of the Andes howl through Tiahuanaco's deserted buildings, it is easy to believe in the Indian legend that the city was built by a race of giants."

At Tiahuanaco there are massive stone stairways that defy simple explanation; they seem to have been built specially to accommodate a race of large people.

There are sections of gigantic stone walls that have been hurled to the ground by what must have been a spectacular earthquake. Wall sections more than a foot thick and 16 by 26 feet in size, have been strewn around like so many toy blocks.

The enormous "Gateway to the Sun" is hewn from a single rock and weighs as much as 10 tons. Other single cut stones here weigh up to 100 tons! Like the other volcanic building stones of this center, it was moved from a quarry at least 60 miles away across rough terrain without wheels and roads! How did they do it?

A Jesuit priest who had apparently interviewed native people of the area in the days of the Spanish conquistadores wrote this amazing account: "the great stones one sees at Tiahuanaco were carried through the air to the sound of a trumpet."[12]

That may sound far-fetched in terms of our experience, but does that mean such a technique is absolutely impossible? The writer of this article in the previously mentioned book must think so. He discounts the priest's early report by writing: "There must have been people even then who were not satisfied with an explanation that invoked the use of magic."

THINK! A hundred years ago, if you tried to explain to a person living then how a Boeing 747 takes off from the runway, you might also be accused of hallucinating or dabbling in witchcraft.

Is there a reasonable explanation?

How Can Massive Stone Blocks Be Moved?

THE ANCIENT RUINS OF BAALBEK, LEBANON

Featured in *National Geographic* magazine in April 1958, these magnificent ruins are labeled another one of those "mysteries."

The prominent 70-foot-high columns seen at Baalbek date back to Roman times. These imposing pillars are amazing. Quarried in Egypt, they were moved across the Mediterranean Sea and then over mountainous regions before being erected here. The engineering to accomplish that feat alone is nothing short of a marvel.

But a glimpse at the foundation stone on which the columns stand is even more surprising. Ancient people before the Romans are responsible for cutting and moving these gigantic blocks from a nearby quarry. The size of these carefully hewn blocks truly boggles the mind.

One of these building stones can be clearly seen where the early engineers moved it only part way out of the quarry. It measures an ***incredible 14 feet, by 16 feet, by 66 feet long.*** It has been estimated to weigh some ***2,000 tons.*** Even with power equipment, our modern engineers have no idea how to move something so immense.

HOW DO MODERN ENGINEERS MOVE HUGE BLOCKS OF STONE?

The National Geographic magazine for May of 1969 featured one of the most ambitious engineering and preservation projects of all time. The great Egyptian monument at Abu Simbel, along the Nile River, had to be moved to avoid being submerged by the rising waters of the newly formed reservoir called Lake Nasser.

The ancient massive statue of Pharoah Ramesses II was carefully cut up and reassembled on higher ground to avoid being drowned by Egypt's Aswan dam project.

The 3,200 year old monument is one of the most striking examples of Ancient Egyptian engineering and art. It was carved into natural rock cliffs beside the Nile. To move it, modern engineers had to cut the towering 67 foot high temple into over 1,000 pieces of 20 to 30 tons each.

By the end of 1969, after 4 1/2 years of work and an expense of forty million dollars the project was complete. The great temple was raised 200 feet to higher ground.

Despite our modern machinery, one can't help but ask: Did the ancients have a simpler way of doing it?

AN AMAZING REPORT FROM TIBET

In a recently published book by authors, Playfair and Hill, *The Cycles of Heaven* includ-ed an incredible report from the Asian country of Tibet. A Swedish aircraft engineer named Henry Kjellson is reported to have witnessed an awesome ceremony conducted by Tibetan priests. At the base of a sheer rock cliff in the mountains, scores of men had gathered to take their parts in the dramatic scene. Groups of them were carefully arranged in a semi-circle, equipped with large suspended drums and others with special trumpets.

As the ceremony proceeded the drum beats and trumpet blasts were directed at the center of the semi-circle in front of the cliff. A four foot block of rock was positioned there. The corporate noise of the assembled instruments must have been deafening. But after a while the heavy chunk of rock (weighing several tons) was seen ascending in the air straight up to the top of the cliff.

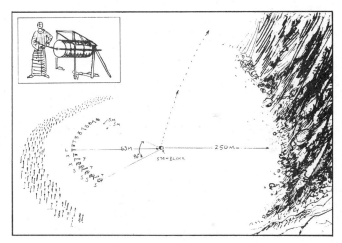

We tend to think of a report like that as being either spiritualistic or false. It just sounds too bizarre to be explainable by natural laws.

But notice a report from the popular *OMNI* magazine in November of 1980. According to the report, NASA scientists have succeeded in using sound waves to levitate pellets of glass or metal!

Could there be realms of technology which we have yet to re-discover? Do you remember Solomon's words? "There is no remembrance of former things. There is nothing new under the sun. It has already been in ancient times before us."

CHINA'S INCREDIBLE FIND

The April 1978 issue of *National Geographic* magazine featured one of the most amazing discoveries of modern archeology. Hidden for years in legend, the unbelievable tomb of China's first emperor has finally been found.

A virtual terra cotta (clay) army, the elaborate burial includes over 6,000 life-size pottery men and horses! Every soldier's face is sculpted with a unique expression.

The excavation will take years to complete.

The total enclosure is over 500 acres! It is called the Spirit City of Ch'in Shih Huang Ti. It dates back to 210 B.C., nearly 2,200 years ago.

The evidence of skilled technology and accomplishment is abundant. Imagine the kilns needed to fire thousands of life-size horses! They are all lined up in battle marching formation, an immense ceremonial parade to escort the emperor into eternity. The whole assembly is completely roofed over and buried. Much of ancient Chinese history will no doubt be filled in as this important discovery is more thoroughly explored.

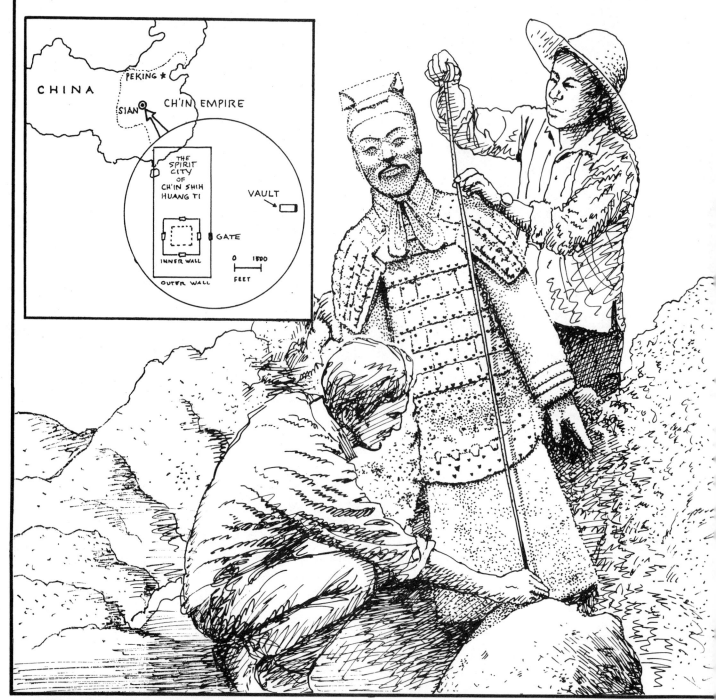

HOW ADVANCED WERE THE ANCIENT ORIENTALS?

The first emperor, whose personal crypt has yet to be unearthed, gives us some indications of the capability of ancient man in the Far East. He accomplished many progressive things and was the first to successfully unify the country. He destroyed the ancient feudal system and centralized the empire.

It was this man, beginning his reign at age 13, who completed the Great Wall of China. The largest man-made structure in the world, China's Great Wall is over 1500 miles long. With 25,000 watchtowers, the twenty-foot-high wall supported a veritable roadway that could have six horsemen abreast march along its pavement.

It was Ch'in Shih Huang Ti who assembled China's first standing army. It may have numbered several million men altogether.

He also codified China's laws, standardized the system of Chinese writing, and built a vast network of public roads and canals.

These accomplishments give us only a little insight into the fabulous ingenuity of mankind. Here in China, as in other ancient centers, we see the propensity of men to manifest that inborn urge to create and dominate.

Endless Discoveries Reveal The Truth About Ancient Man

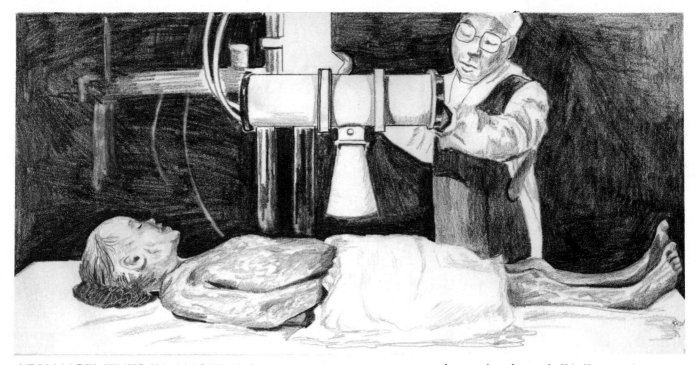

ADVANCEMENTS IN ANCIENT CHINA

Another remarkable burial, this time of a noblewoman, was featured in the May 1974 issue of *National Geographic* magazine as "A Lady From China's Past."

The elaborate burial was like an underground building, carefully engineered with protective layers of charcoal and white clay. Several ornately decorated caskets were found, each inside the other.

When the Chinese specialists opened the innermost casket they were in for the surprise of their lives. Under 20 layers of fine silken wrappings the astonished scientists discovered the woman's body was almost perfectly preserved! Her skin is still pliable, her hair is still firmly rooted in her scalp. Modern X-ray technicians can even identify the organs of her body, and the menu of her last meal eaten the day she died!

How did they preserve this corpse so well 2,100 years ago? Advanced knowledge would have been necessary to halt natural decay processes so immediately and for such a long time. In our modern age we are just beginning to understand such preservation by observing what the ancients did.

Found in the tomb with this woman's body were a number of other skillfully made artifacts. Each of them further opens our understanding of the fact that these ancient people were quite technologically advanced in many ways.

An ornately decorated cosmetics case accompanied this high-class socialite. In the case was a silk scarf and other 20th-century conveniences like mitts, hairpiece, face power, rouge, comb, and brushes.

Also found among the grave goods were 180 pieces of exquisite lacquer ware jars and table service items. These lightweight containers are highly crafted and artistically decorated. In ancient China these acid-resistant vessels were esteemed ten times more costly than similar bronze objects.

Even more surprising was the discovery of printed cloth! Fine silks numbering some 50 lengths, comprise the most lavish cache of ancient fabrics ever found. When you look closely at the intricate repeated designs on the cloth you realize they are not painted on; they are printed. And keep in mind that this was done over 2,100 years ago. Maybe the craft was just re-invented by Hans Gutenburg in about 1450 in Germany. Again, this is just another indication of the truth of Solomon's words of 3,000 years ago: "There is nothing new under the sun."

NOTHING NEW UNDER THE SUN

As you continue to read news reports of astonishing finds produced by ancient civilizations, the key to the mystery becomes clearer.

Should we be surprised to learn of a glass lens produced in ancient Egypt?

In 1898 another odd discovery was made among the treasures of ancient Egypt. The little wooden carving was stashed away in the archives for years and simply labeled as another image of a bird. But recently it has drawn special attention. Because of its unique vertical tail fin and streamlined wing design it was put through some tests and found to be aerodynamically ideal for mechanized flight.[13] Five years before the Wright brothers flew at Kittyhawk, this discovery meant almost nothing. Now we are forced to deal with the possibility that ancient people were aware of, and may even have used, flying machines.

Some ancient human skulls reveal that cranial surgery was performed with at least some success on the Inca people of ancient Central America. We know the patients lived because the cranial bone tissue shows a partial refilling of the opening through the natural process of healing.

In modern time, dentists have learned to use the lost wax method for making tooth crowns. But man-made tooth crowns were also fashioned by Aztec dentists of ancient Mexico!

Traditions of the Mayans of Central America and Mexico insist that phenomenal things were done in the realm of communications. A 1,000 pound slab of solid crystalline rock rests in a special high tower in the mysterious ancient city of Palenque. The priestly rulers of this vanished civilization are said to have come here to transmit and receive audible messages from great distances away.

DID ANCIENT COMMUNICATION SPECIALISTS USE MODERN CRYSTAL TECHNOLOGY?

There are a number of indicators both from the Bible and from the physical findings of ancient cultures that rock crystals were used to communicate through the air. Though this idea may seem to be granting too much knowledge to early man, we must be honest enough to continue exploring for verifications of Solomon's statement in Ecclesiastes chapter one. [This subject will be treated in graphic detail in an upcoming supplement to this book. For further information write to the address on the title page.]

HOW COULD ANCIENT MAN BE SO ADVANCED?

No matter where we travel in the world, when we encounter ancient civilizations the story is always the same. Their tell-tale physical remains continue to baffle our modern science.

Architecture, engineering, specialized skills, arts and industry continue to show up in the most remote historical times. The cradle of civilization reveals no primitive past preceding the great cultures of Egypt, Sumeria, and Babylon. These cultures produced marvels of human ingenuity.

Yet why is it that cleverly imagined artwork presistently popularizes the notion that ancient men were primitive animals? Men's vain theories stubbornly resist the clear fact of all recorded history:

MAN HAS ALWAYS BEEN 100% HUMAN!

WHY IS MAN SO INGENIOUSLY CAPABLE?

No one recognized the answer more profoundly than King David of 3,000 years ago:

"You (Almighty God) have made man a little lower than the angels, and You have crowned him with glory and honor." (Psalm 8:5)

References For Section Four

Many of the references for material in this section have been cited within the text. Much of the information has been reported thoroughly in books and articles dealing with ancient civilization mysteries. Sometimes it is difficult to separate special data from what is called "general knowledge." It is not our intent to overburden you with long lists of sources for all the facts. Much of the general information can be found in the selected reading titles listed below.

1. "A Lady From China's Past," National Geographic magazine, May 1974.
2. Ibid.
3. Megertchian, Korium, "Metallurgic Factory," Bible-Science Newsletter, Five Minutes section Feb. 1978, p. 3.
4. "Ancient Electroplating," 1933 Annual Log, Scientific American Publishing Co., New York, 1933, p. 85.
5. VonDaniken, Erich, In Search Of Ancient Gods, G.P. Putnam's Sons, New York, 1973, p. 173.
6. Fell, Barry, America B.C., Simon and Schuster, New York, 1976, p. 21.
7. Polynesians, National Geographic magazine, December 1974.
8. The World's Last Mysteries, Readers Digest Association, Inc., Pleasantville, New York, 1967, p. 94.
9. Schwalb, Harry M., "Electric Batteries Of 2,000 Years Ago," Science Digest, v.41: p. 17-19, April 1957.
10. Ibid., "Ancient Electroplating."
11. The Mayan, National Geographic Magazine, December 1975, v.148:6, p. 783.
12. Ibid., The World's Last Mysteries, p. 138.
13. "The Little Wooden Airplane," Pursuit, v.5:p.88, 1972. in Ancient Man: A Handbook Of Puzzling Artifacts, by Willian R. Corliss. (see below).

Selected Reading

Gaverluk, Emil, Did Genesis Man Conquer Space?, Thomas Nelson, Nashville, 1974.

Richardson, Don, Eternity In Their Hearts, Regal Books, Ventura, 1981.

Corliss, William R., Ancient Man: A Handbook Of Puzzling Artifacts, The Sourcebook Project, Box 107, Glen Arm, MD 21057, 1978.

Kang & Nelson, Ethel, The Discovery of Genesis (in the Chinese language), Concordia, St. Louis, 1979.

INDEX

INDEX

Dennis R. Petersen, B.S., M.A.

Since the mid-1960's the author has been involved in many research and instructional pursuits. His life-time interest in natural sciences and history has resulted in a wide scope of research from which to draw in his preparation of this book. After working professionally with several museums in the United States, he was uniquely led by God to Canada, where he received his formal Bible education.

Since 1974 he has taught Old Testament subjects at a Canadian Bible college; pastored for five years; and presented seminars on science and the Bible in many cities of Canada and the United States.

As the founder and president of the Creation Resource Foundation, Dennis and his lovely wife, Viola, make their home with their four fine children in El Dorado County, California. He has been an active participant and advisor in the local church there since they relocated from Canada in 1981.

A Dynamic Multi-Media Seminar Exploring the Wonders of Our World and Hidden Truths of the Bible

The very foundations of Biblical faith are being deliberately undermined, and it's happening in the guise of what appears to be science!

ARE THE BIBLE ACCOUNTS OF CREATION, THE FLOOD, AND OTHER RELATED ISSUES REALLY VERIFIED BY HONEST INVESTIGATION?

This extraordinary seminar is a life changer for the entire family!
COLORFUL MULTI-MEDIA THROUGHOUT

This timely and unique Bible-based ministry presents surprising and colorful insights on subjects like:

- **Mysteries of Ancient Civilizations**
- **Dinosaurs and Cavemen**
- **Fossils and Evolution**
- **Mysteries of Life**
- **Extra-terrestrials**
- **Missing Links**
- **The Perfect Earth**

"… the case he (Dennis Petersen) made for creation was quite impressive. … I am happy to recommend this seminar to others, believing that the case for creation is one that is vitally important in our age and that Christians must become better informed on the issue."

Dr. D. James Kennedy
Coral Ridge Presbyterian Church
Fort Lauderdale, FL

"… Excellent academically and scientifically!
… Excellent scriptually and theologically!
… Excellent in meekness and kindness!
You need to have this Creation Seminar in your church."

Pastor Ernie J. Gruen
Full Faith Church of Love
Shawnee, KS

For further information on how your church or organization can schedule a seminar in your city contact:

Creation Resource Foundation
P.O. Box 570 • El Dorado, CA 95623 • (916) 626-4447

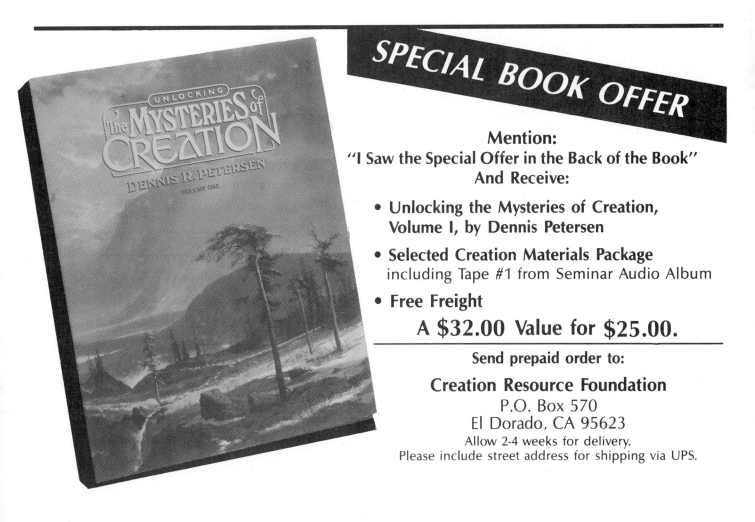